LIVING BY SURPRISE

LIVING BY SURPRISE

*A Christian Response
to the Ecological Crisis*

WOODY BARTLETT

Paulist Press
New York/Mahwah, N.J.

Book design by Bookcomp, Inc.
Cover design by Cheryl Finbow

Library of Congress Cataloging-in-Publication Data

Bartlett, Woody.
Living by surprise : a Christian response to the ecological crisis /
Woody Bartlett.
p. cm.
Includes bibliographical references (p.   ) and index.
ISBN 0-8091-4142-6
1. Ecology—Religious aspects—Christianity.   I. Title.
BT695.5 .B376 2003
261.8'362—dc21
2003001686

Published by Paulist Press
997 Macarthur Boulevard
Mahwah, New Jersey 07430

www.paulistpress.com

Printed and bound in the
United States of America

To
Samantha and Avery,
Zachary and Anna,
their children
and
their children's children

CONTENTS

Introduction

1

CHAPTER 1

Endangered

5

CHAPTER 2

The Big Bang Beginning

30

CHAPTER 3

Explosion of Life

53

CHAPTER 4

Humanity

81

CHAPTER 5

So Here We Are

104

CHAPTER 6
And There We Go
131

Annotated Bibliography
157

Appendix 1: The Lineage of Humanity
160

Appendix 2: Environmental Information
161

Index
163

Introduction

PICTURE YOURSELF SITTING in the woods, propped up against a tree and feeling the warm sun on your face. It is a deeply satisfying experience. Insects crawl through the grass and leaves. Over to the side, a squirrel bustles. A butterfly wanders by, and a bird sings in a nearby tree. The Earth is going about its business of living as it has for billions of years. All seems well with the world—and it is.

But a discordant sound rustles at the back of your mind. There is something not right. What is increasingly not right is the effect you and your species—humanity—have on the Earth. That discordance is hard to discern on a warm spring day in the woods, but the early warning signals are being heard. And people are starting to respond.

Propped up against that tree, you cannot easily understand that Western humanity has launched an assault for control over that scene around you. In general, trees are seen for their economic value. Land is understood for what can be built on it. Development is prized above every other consideration. And it is the cultural norm for people to consider themselves as absolute owners of the land, able to do with it whatever they desire.

The effect is an increasing discordance on the Earth. Weather patterns are changing. Land, air, and water are being poisoned. Trees are being felled. Animal life is suffering, including the disappearance of entire species. All is not well. A new way for humans to live on the Earth is needed.

When we seek an understanding of why we are here and how we might restore some balance on the Earth, it is important that we look to the great thinkers of the past, to science and to our faith for clues to that understanding. In doing that, this book presents four Dynamics. These Dynamics are then used to describe the nature of the creation and the nature of God's action in and through that creation. They give us important clues about how to live on the Earth in order to move through the ecological crisis.

The first chapter explores how we got into this predicament and what it will take to recover. The next two chapters tell the story of the universe, starting at the Big Bang beginning and moving through the story of the stars, the story of the Earth and organic life on the Earth.

Since humanity is at the center of the present ecological crisis, the fourth chapter explores this big brain experiment called *Homo sapiens*. Why are we like we are? How did we get this way?

The fifth chapter looks at the present ecological crisis, how it relates to the four Dynamics, and how the church reflects that crisis.

The final chapter explores ways to follow the four Dynamics so that we might come to a more harmonious relationship with the Earth. And it looks to the Christian faith for the power to bring about that harmony between humanity and the creation.

Story form is used because all peoples know who they are through their stories. Persecuted minorities spend a lot of time on their histories. Majority cultures know who they are through ritual and special holidays and the celebrations of their heritage. It is important to know where we come from. So it is essential, in our time, for us humans to know our story in its broadest sweep, from the beginning of the universe to the present. This story can then provide guidance and power for life in the future.

In order to complete the healing, we must also unify our most personal experiences with our understanding of the whole cosmos. So each chapter starts with a concrete personal experience and uses that experience to point to the most universal workings of the creation. It is a way to unify the micro and the macro, the smallest and the largest, the particular and the global.

This whole book could be written in a more academically formal, historical, scientific, or theological way—and perhaps it should be. But not by me. Instead, I have chosen the way that is most comfortable to me, telling the story the way I as a parish priest might tell it to my people. It is in lay language as much as is possible. After all, it is our story.

It would be ungrateful to proceed without acknowledging the considerable debt I owe to others for helping me get to this place. First and foremost, there is that great French thinker, Pierre Teilhard de Chardin, who showed me for the first time that it was possible to join science and religion together in a common system. His *The Phenomenon of Man*[1] laid the groundwork.

Then I must mention his modern-day interpreter, Dr. Thomas Berry, and his eloquent disciple, Sister Miriam McGillis. Without them my imagination would not have been stimulated to push as far as I have. The immediate inspiration for the project came from Dr. Brian Swimme, whose stories of the inner life of the universe are captivating.

There were, of course, books and books. The thinking of Niles Eldredge and Stephen Jay Gould, James Lovelock, Lynn Margulis and Elisabet Sahtouris, Peter Russell, Ernest Becker, Stephen W. Hawking, Gro Harlem Brundtland, and many others—read and unread—has been invaluable. Without their stretching, there would be nothing.

And there were those who assisted in the writing by reading, rereading, and critiquing the text. My wife, Carol Bartlett, and Dr. John Westerhoff were most helpful, as were Dr. Charles Wharton, Dr. Danny Martin, and Olin Ivey. Ms. Nell Jones assisted greatly with patient editing and enthusiastic encouragement. Without all of them, the work would not be complete.

Then there is the rich heritage of the Christian faith, with its understandings of the ways of God and humans. It is my tradition, and I am steeped in it and thankful for the knowledge and context it has provided.

Finally, however, the greatest teacher of all is the Earth upon which we all live. Out of the Earth comes the direction in which we

---

[1] New York: Harper Torchbooks, Harper and Row, Publishers, 1959.

must go. All of us, if we would survive, must be its students. We must listen to its rhythms, watch its cadences, catch its spirit. In the Earth is contained the wisdom of the ages. And through the Earth speaks the Lord God, the Creator of it all.

CHAPTER 1

Endangered

SURPRISE! RIGHT AT the heart of the universe, there is surprise. Out of that heart spot, from time to time, surprise jumps up into our lives and gives us a start, sometimes by knocking us around and down, sometimes by just brushing up against us.

Take that day in San Diego. I was there for a planning meeting preceding a conference, but I couldn't find the meeting. As it turned out, the meeting and the conference were not the real reasons I was in San Diego. I was really there to have my life changed by surprise.

So there I was in San Diego with a day to myself, for something. What do you do with a free day in San Diego? You go to the zoo, of course ...

I'd never been to the San Diego Zoo before. It is a beautiful place, with many exotic creatures exhibited in their natural habitats. There are just a few cages and bars. Mostly, it is animals with humans looking at them like so many honeybees around the first blossoms of spring.

As I wandered along the pathways, my curiosity was piqued by a red triangle with a white *E* in it at the top right corner of many of the exhibit markers. Looking in my guidebook, I was startled to see that it meant that that species was "Endangered." Then I noticed the explanation off to the side of each of these displays that told how and why that particular species was so vulnerable. Almost all mentioned the presence of humans as the prime cause. The red triangles then took on a blood-red tone. At the thought of the loss of so many gorgeous examples of God's creative exuberance, my eyes misted and the beauty of the spring day took on a more melancholy hue.

Finishing my tour of the zoo in the early afternoon, I went across the park to the Natural History Museum. It had the usual array of dinosaur bones and stuffed raccoons. But in one corner was a special exhibit on extinction. With the red triangles still in mind, I headed for the entrance. Again were chronicled the numbers of species that were becoming extinct in our day. It was an all too familiar litany. The exhibit pointed to the influence or presence of man, *Homo sapiens*, as the cause of 99 percent of the extinctions!

Even the Laplanders from the Arctic sections of Scandinavia and Russia were in the display. There they stood, dressed in their colorful garb just like I fondly remembered them from sixth-grade geography. The caption said they subsist on reindeer meat and that their reindeer got a lethal dose of radiation during the Chernobyl disaster. Radioactive fallout, eaten by the reindeer, would likely be passed on to the Laplanders, affecting their genetic codes and possibly ending in the demise of their tribe. Things really did get visceral then.

Finally, I reached the end of the exhibit and was faced with a final chart whose message came crashing in on me. It read:

Rate of Extinction
Before Man—1 species per 1000 years
During the age of dinosaurs—100 species per year
At the present time—1000 species per year
At the present rate—in 20 years—several species per hour
When humanity?

I was stunned at what I was seeing. We are in a period of extinction that is much, much harsher than that which had once wiped the mighty dinosaurs off the face of the Earth. At that time, up to 90 percent of the existing species had disappeared. If it happened to them, it could surely happen to us. I became haunted by the image of a cage at the zoo with humans milling around in it. The marker would say *"Homo sapiens"* and it had a red triangle with a white *E* on it— Endangered!

I stood there blinded by tears as St. Paul must have stood on the Damascus Road centuries ago. I had known in some dull-witted way that we were harming our environment in dangerous ways. But this? This was much more than I had imagined. I groped for the exit and encountered one last sign.

It read:

If all the beasts were gone, men would die from great loneliness of spirit, for whatever happens to the beasts also happens to the man. All things are connected. Whatever befalls the Earth befalls the sons of the Earth.

Chief Seattle (Washington)
to the President of the U.S., 1855

ALL I COULD picture was a great crash of life, carrying along with it me and my kind to destruction. I thought of my faith. If Christ came to save us, then wasn't this what salvation should be about? But all my theology concerned personal salvation; it seemed to have no relation to the extinction of the human species. What made a species live and what brought it to extinction? Did the life and message of Jesus relate to this issue at all? Ever since, my life and thoughts have been driven by these questions.

When I got home and told the story, someone pointed out that the new temporary 25-cent stamp from the U.S. Post Office had a picture of the Earth on it and was marked with a big white *E!*

THE CRISIS

I hadn't previously known much about the environmental crisis and so had to play catch-up. Many others have had to also. These past few years we have become familiar with the issues of ecology, pollution, and population growth. We now know the ozone layer in the upper stratosphere is measurably thinning. We know the so-called greenhouse gases continue to accumulate in the atmosphere and contribute to the global warming that will probably severely change our weather patterns. We know we are losing our topsoil, adding toxic wastes to our drinking water and land, cutting down our trees, losing our rain forests, and generally stressing the Earth in dangerous ways.

People are responding. Recycling has become standard practice in many communities. Curbside recycling is common in metropolitan areas. Most schools and many shopping areas also have recycling bins in the back of their buildings. America Recycles Day has been held the past few years, with the 1998 day having 4143 confirmed events across the country. During those events 2.1 million pledges were collected. Recycling is becoming a stable part of the modern scene. Today many of us know that something is deeply wrong with the environment. Just as Hamlet knew that "time is out of joint," so many people in this country know that we have a giant problem.

We are not the only country to come to this conclusion. In September 1987, most of the nations of the world agreed to eliminate the manufacture of chlorofluorocarbons (CFCs), a particular organic compound used largely in refrigerants and air conditioners. Loose CFCs rise through the atmosphere, combine with free ozone, and endanger the ozone layer that protects the Earth from dangerous radiation from space. The Montreal Protocol, controlling the manufacture of CFCs, was speedily agreed upon and has been updated four times since, a remarkable performance. Five years later, the largest summit ever held, the Earth Summit, took place in Rio de Janeiro in June 1992. Over 60,000 people from around the world attended it. Five years after that, in 1997, most of the nations of the world entered into a treaty on global warming—the Kyoto Accords. Although the treaty has not yet been widely ratified, it shows that the people of the world know something is deeply wrong.

It seems we want to do what we have always done: rush out and fix the problem. Get an alternative to CFCs; reduce carbon dioxide emissions into the atmosphere; clean up the toxic waste dumps; get better sewage treatment and industrial waste treatment plants. To be sure, all of these responses are necessary and important.

The issue, though, runs much deeper than improved technology. Albert Einstein, who had a universal grasp of things, said in one way or another on many different occasions during his life that you can never solve a big problem by addressing it with the same assumptions that you had when you got into the problem. If you are driving in a strange city while using an outdated map and you get lost, it's pretty difficult to find your way with the old map. You need a new, more accurate map. If you build a house using specifications calling for a certain kind of nail and the house starts coming apart because the nails are too short, you will never fix the house using more short nails. You need new specifications calling for new hardware—including longer nails.

We all want to fix our environmental problems with new and improved technology. But a Roman Catholic nun, Miriam McGillis, helped me to see that what is needed is a basic shift in the way we think, in our assumptions. Technology is fine as far as it goes, but what we need now is a new map. What we need are new specifications for the house. This book will provide the outline for a new way of thinking so that humanity can live easily within the greater Earth system.

The Ways We Think—Cosmology

In talking about our assumptions regarding, and our understanding of, the creation, we are talking about cosmology. *Cosmology* means knowledge of the cosmos or the world. *Webster's* defines it as the "theory and philosophy of the nature and principles of the universe." Though a deep and vast subject, we have to understand it in order to go deeper into the environmental issue.

Everyone has a cosmology. Every culture has a universe story, which gives meaning to reality. It informs us of the way we should relate to the creation. It attempts to shed light on the deepest mysteries that confront us: the meaning of life, our role in life, our origins, the meaning of birth and death, the meaning of pain and suffering, the nature and function of God.

Usually, we are incredibly unaware of our cosmology. It is so ingrained, such a matter of common sense, so second nature that it is hardly ever verbalized. Our cosmology is just taken for granted.

But people from other cultures stumbling into our culture become quite aware of some of our assumptions. Since they do not hold the same assumptions, ours stand out like a reflector sign before head-lights at night. For instance, have you ever traveled in a town where the street signs only mark the cross streets and never tell you the name of the main street? Everyone there knows the name of the main street—except you. You have to go to the end of the street or find where it intersects with an even bigger street in order to learn its name.

At the root of every culture lie some given assumptions that are passed on from generation to generation, as if by osmosis. We learn them from our parents. We seem to pick them up out of the air we breathe. We learn them from the talk on the playground, from what we see or don't see on television, from what is discussed or avoided at the dinner table. Unless our families are exceptional, we learn that sex is a very private, even dangerous, thing. We learn that God is in Heaven and that Heaven is up. We learn that when we are finished with something we can throw it away and we will never have to be bothered with it again. Those lessons profoundly affect the way we do things—sometimes with drastic consequences.

A story from World War II illustrates the clash of cosmologies. It seems that some lend-lease tractors were sent to remote villages in India. The people were good farmers but uneducated in the ways of the West. In fact, they had never seen machinery like this before. They were in awe of it.

So technicians were sent with the new tractors to explain them and teach the people how to use them. Despite the vast cultural dis-tance the farmers had to travel to master these tractors, they learned quickly and became quite adept at using their new tractors.

The technicians also painstakingly taught the farmers the neces-sary maintenance procedures—where they would have to grease the tractors, how they were to change the oil and how often. The farm-ers learned that equally well. And so the technicians went home.

The farmers were ecstatic with their new power; it was awesome to them. So in addition to the regular maintenance, they performed their native rituals on the tractors. They danced and chanted to keep the machines performing well. After all, their cosmology said that everything works well when it is surrounded with the right spirits and those spirits are satisfied.

Things went along marvelously. The tractors worked wonderfully. The farmers were pleased that their rituals worked so well. The steady roar of the tractors was continuing confirmation. Though they performed the oil changing and other maintenance procedures at first, as time passed the less they saw the necessity of that. They just couldn't understand what difference it would make. Finally, they stopped changing oil and doing maintenance and just did their native rituals. The tractors eventually broke down.

The farmers' cosmology didn't have a place for the mechanical side of tractors. Western technology didn't fit into their grand scheme of things. Their thinking was not mechanical, but spiritual. The maintenance of the machines, finally, had no place in their universe and was ignored. That is not to say that the farmers were inferior or dumb; they just had another point of reference. They had another way of thinking, and that way of thinking wasn't just an idle curiosity. It destroyed the tractors. At the beginning, they should have placed a sign on the tractors reading, "Endangered."

A culture's cosmology is so engrained that it is not easy to change; it often takes a severe crisis within the culture to do so. But it can be done. New cultural myths can be born. They drift up out of the psyche of a culture to explain and give meaning to the new conditions in which the culture finds itself. Such a time could surely be our time.

As we look at the environment becoming less and less hospitable to our human species because our species is becoming less and less kind to it, we must also look carefully and deeply at our way of thinking. We must ask questions about the adequacy of our culture's cosmology to the task of maintaining sustainable life for the human race. Hopefully, a new, more useful cosmology will emerge from this introspection.

Before going further, let's consider a couple of cosmologies so that we can get a better idea of all of this. First, Native Americans, also known as American Indians, have a cosmology that is in remarkable contrast to that of Western culture. Much like the Asian Indians with the tractors, they see the spirit in everything. The forests, the animals, the wind and rain all reflect and reveal the spirit. Indians see themselves as part of a community of life that is far bigger than they are. In fact, many Native American hunting rituals reflect an understanding that the hunter is a member of a larger community of nature of which the animal is also a member. At the moment of the kill, the animal is offering itself up to the hunter as a gift. Therefore, the hunter gives a ritual gift in return. In the ritual he reminds himself that the hunt is a part of the larger relationship of animal and hunter. The hunter thanks the animal for his sacrifice so that the animal species will not be offended.[2] This is also true of vegetable nature. Indians will often make an offering of tobacco to the ground in exchange for the fruits or vegetables that they pick. Everything is personal and relational.

For the Native American, suffering and death are also part of the natural order. While they are feared, they are also accepted as part of the way things are. The purpose of life is to bring one's individual spirit into participation with the rest of life—to be in a community of giving and taking, learning, and sharing.

One Native American at a conference I attended told a story about his own early-adolescent son who was growing up in very Western ways. The father did not want him to miss the treasures of the Indian culture. So he took him out into the woods one day. They traveled for miles and finally reached a running brook high in the mountains. He told his son to sit on a large rock next to the brook and see what he could learn. The father went away and returned in two hours to ask his son what he had learned.

"Nothing," was the reply.

So the father went off for another two hours and returned again. Again, the son had learned nothing. This happened two more times, after which the son finally said that he had learned something.

"What?" asked the father.

"This rock is very patient," replied the son.

Western cosmology, which informs our lives, is quite different. In our mind, the Divine is not an immanent presence living in and through all of the creatures on the Earth. Rather, the Divine is transcendent, living in the realm of the spirit, in Heaven. The task of humans, in this cosmology, is to escape the physical realm through prayer and salvation, in order to meet God in Heaven.

---

[2]Joseph Campbell, *The Power of Myth* (New York: Doubleday, 1988), p. 74.

In Western cosmology, the physical universe is devoid of spirit; in fact, it is often opposed to spirit. Death and suffering are physical imperfections, without spirit and not a part of life as it is meant to be. The implication is that there was a time when death and suffering did not exist, a time of innocence. This is described in the Bible as the Garden of Eden, ended by the Fall. Again, in the biblical understanding, there will be a time in the future, the final millennium, when death and suffering will be no more. The result of all of this is that we are free now to probe and manipulate the natural world and all of its creatures in seeking to bring about the millennial end of suffering and death.

This Western cosmology has allowed us to encircle the globe with transportation and communication. It has allowed us to go to the moon and plumb the depths of the oceans. It has allowed us to eliminate some deadly scourges and control others. We have held this cosmology for a long time, but there are increasing breakdowns. Warning signs are beginning to flash that say to us that the old cosmology is no longer working so well. We are becoming an endangered species. It is happening because of the way we think.

How We Got Here

In the months following my encounter with surprise at the San Diego Zoo, I read books and articles, attended conferences and seminars, and woke up many mornings, thinking, "How did we get here, and how do we get out?"

I've come to some beginning understandings of Western cosmology and its role in our dilemma. Our Judeo-Christian heritage is surely one of the strongest strands in our culture. So we turn to it first. In the Genesis accounts of the creation, it is clear that God created the universe from nothing, *ex nihilo* as the learned theologians say it. In this cosmology, the universe is thus clearly outside of God. It is of lesser stature. It did not emanate from God as if the creation were itself God. It was created as a potter might create a pot. It was formed and shaped by God. In fact, the human race, created in the image of God (Gen. 1:26), was given dominion over all of the creatures on the Earth (1:26) and told to multiply, fill the Earth, and subdue it (1:28).

In almost every conversation I've had about humanity and the creation, the question of dominion has come up. So let's look at that more closely to see what it might really mean. It is helpful to start by looking freshly at the image of God from within the context of the story. What is God's image in this story? Clearly, God is lavishly creating the heavens and Earth. The image is one of exuberant generosity, of creative extravagance, of purposeful conception. Beyond that exuberant creativity, God also celebrates what has been done. "And God saw that it was good" (1:4, 10, 12, 18, 21, 25, 31). So God created humanity in God's image to be generously creative and to celebrate the results of that creativity!

Dominion clearly means that humanity has power. Power is the translation of the original Hebrew word. But to what end are we given power? What are we to do with this power? If we were created in the image of God, then dominion (power) is given so that we might be co-creators with God. It is given for the sake of the creation. The focus in this story is on the well-being of the creation.

The one slight exception to being generously creative for the sake of the creation is stated in Genesis 1:29. There it says that God gave every plant and tree to humanity: they were to have them for food. It does not say that dominion is given so that humans might be rich or comfortable or air-conditioned. The only way we should use our power for ourselves is to eat, with the hint that we should be vegetarians at that!

Then comes the Fall as recorded in the second chapter of Genesis. Adam and Eve are expelled from the Garden of Eden and told that woman would now conceive children in great pain and man would have to struggle with the earth in sweat and tears to obtain food. Humanity still has power but, now corrupted, centers the power on itself rather than on the creation. It is now characterized more by domination than creation with celebration. In our post-Fall condition, the focus has changed from the creation to ourselves. We have warped our gift of dominion to mean domination, license to exploit, permission to control.

The more we try to control the creation, however, the more pain we suffer. We were made for love, not greed. We were made to be co-creators with God, not exploiters of God's creation. We were made for celebration, not self-glorification. We have twisted the purpose of the creation to our own ends rather than God's ends.

The other strong influence on our cultural heritage comes from the ancient Greeks through the Romans. They had many strengths but one of the most important was clearly their prowess at analyzing and organizing. They had the ability, different from most all their neighbors, to break down any subject into its separate components and then organize those components rationally. Without the influence of the Greeks we wouldn't be able to make outlines with categories, subcategories, and sub-subcategories that are familiar to schoolchildren. But without that ability to organize, we wouldn't have any kinds of systems—manufacturing systems that produce cars and televisions, defense systems that keep us free, organizations such as major league sports and large concerts. To be sure, we also wouldn't have school systems. Without the ability to analyze and organize, ours would be a radically different society.

Using this ability to organize, the Greek and Roman civilizations flourished and dominated the world for centuries. Often their technical skills were used to develop great cities. One such city was Ephesus in what is modern-day Turkey. It became a booming port city, but because of deforestation and exploitive farming methods, it was turned into a wasteland with denuded land and a silted-up harbor. The small size of the human population in those days made it possible to be very destructive and still move on to other places without contributing to a global crisis.

Some early figures in Western culture had a wonderful understanding of the proper relationship between God, humanity, and nature. St. Francis of Assisi (1182–1226) was a devout man of God, looking to God daily for guidance for his life, particularly in his deep concern for the poor. But St. Francis was more than that. He embraced nature and all of its creatures as well. He displayed an immediacy that addressed Brother Sun, Sister Moon, a love of and respect for animals, a gentleness of spirit that walked lightly and simply on this Earth. He was a remarkable testimony to the possibility of faithful humans living in an ecologically sustainable relationship with the creation. We might call his viewpoint an integrated cosmology.

Thomas Berry, a Passionist priest, a cultural historian, and one of the most articulate modern prophets discussing the place of humanity within the Earth system, pointed out in a lecture I heard him give in New York in 1990 that in those days humans recognized two great authorities. They read the Book of the Scriptures to see what God said, and the Book of Nature (that is, the world around them) to see what nature, the creation, said about what they should do.

But within a hundred years of St. Francis, a disaster occurred that changed the shape of Western culture forever. Around 1350, a plague called the Black Death struck Europe. Up to 50 percent of the people died in some areas, and the total population of Europe declined by at least a third. Cities were thrown into utter confusion. People wailed about a supposedly merciful God allowing this to happen. It was a devastating trauma. Mother Nature struck cruelly and silently in a way that no one understood.

In his book, *The Dream of the Earth*,[3] Berry suggests two broad cultural responses to this trauma (pp. 125–127). One emphasized a religious redemption out of this tragic world. If life was to be this dangerous, let us find our home in Heaven in the next life. A religious concentration on sin, salvation, and the afterlife emerged. This is quite a contrast to St. Francis and his celebration of the community of all creatures here on the Earth.

The other response was to search for a means to prevent this from happening again. Society set out to understand and control the physical world in ways that it had only briefly glimpsed before. Thus emerged the scientific and technological, secular culture that so dominates our day.

The Black Death was a trauma that shook the prevailing cosmology of the fourteenth century, setting in motion cultural currents that resulted in the separation of science and religion. Science with all its knowledge, and religion with its values, went their separate ways, sometimes with a shrug, and sometimes yelling at each other at the

---

[3]San Francisco: Sierra Club Books, 1988.

tops of their lungs. Two cosmologies fought for the heart of their society. The results, over the ensuing centuries, have been devastating.

THREE MILESTONE THINKERS

Three men of the fifteenth and sixteenth centuries demonstrate the basis for our conflicting modern Western cosmologies. They are Francis Bacon, René Descartes, and John Calvin. The time in which they lived was turbulent. This was the age of discovery, with Columbus, Cabot, and Magellan charting unknown lands across the seas. The arts were experiencing a revolution called the Renaissance, resulting in the whole orientation of Western art shifting from an eternal setting to an Earth-based, human setting. (Just compare the gold backgrounds and stylized figures in pre-Renaissance art with the work of Leonardo de Vinci and Michelangelo and Raphael. In the latter, the settings are in this world.) Like art, religion was in upheaval with the Protestant Reformation. Copernicus and, at a later date, Galileo revolutionized astronomy and the perceived place of humankind at the center of the universe.

The whole process of Western thinking was changing. Up to that time, philosophers had taken ancient truths and applied them to present conditions. The authority for truth resided in the past, with Plato, Aristotle, and others. But this incredible age of discovery changed all that.

Present discovery became the authority, and new truths had to be derived from these new discoveries. It was a whole new approach to the world, and it resulted in two new cosmologies that are present yet today.

The English politician and scientist Francis Bacon (1561–1626) made a crucial contribution. Seeking to shake up the philosophical establishment, Bacon asserted that the central question was not *why* things worked, but *how* they worked. So he outlined a program of inquiry into the way nature works. There were to be no preconceived ideas about the outcome. The observer was to be strictly neutral, collecting data from experiments and deriving laws of nature from the results. Bacon began the scientific method and changed the course of history. It was the beginning of a new cosmology that assumed that nature could be analyzed, categorized, and, by implication, controlled. The human observer took a position outside of nature and merely reported the truth as revealed by nature. To Bacon's credit, he always insisted that humans operate from a position of charity toward nature in their scientific inquiries. Unfortunately, that has been too often forgotten as modern humans using science are often motivated by greed and the lust for power.

The second person who changed our way of thinking was René Descartes (1596–1650), a French philosopher, scientist, and mathematician who lived almost at the same time as Bacon and who struggled with some of the same academic bankruptcy.

Descartes, in searching for the ultimate reality, argued that he could not trust any information derived through his senses because it was always possible that he was dreaming or otherwise misled. The only thing to be trusted was reason. This assertion was captured in one of the most famous statements in all of philosophy, "I think, therefore I am" (*cogito ergo sum*). It led to a sharp distinction between the mind (reason) and the body (nature). Furthermore, Descartes saw the body as inferior to the mind. Since Descartes was a mathematician, he saw the body and the rest of nature operating by strict mathematical laws. The entire universe was operating mechanically, like a very complex clock.

This body of thought strongly undergirded the dualism that had haunted Western philosophy since the time of the Greeks and has plagued it ever since. We will have much more to say about this later.

So we see two seventeenth-century scientific thinkers describing humans as separate from nature (in Descartes' case, the mind as separate from nature). They argued that the whole universe operates by set laws that can be understood. And if they can be understood, they can be controlled. We might call this way of thinking the domination cosmology.

Meanwhile, what about the religion of the West? Where was it after the great split following the Black Death? We can look to perhaps the most influential theologian of the era, John Calvin (1509–1564), for some clues. Though somewhat earlier than Bacon or Descartes, Calvin was also involved in reform. This reform was against the principles of the Roman Catholic Church that involved the church theologically and practically in the detailed affairs of this world and anchored its authority in the remote past.

Calvin's thought centered around the absolute sovereignty of God and, therefore, the immediate authority of God. The predestination of every human being to Heaven or Hell logically followed. Much of his attention went to the state of the human being in the afterlife. Rather than seeking control of this world, Calvin gave it to God and let the spiritual chips fall where they might. Later followers of Calvin have insisted that this eternal place in the afterlife must be a direct result of a conversion experience, usually quite dramatic and emotional, in which the believer was "saved." So in this understanding, all of the attention of the church and Christians was to prepare for the afterlife. Another new cosmology was emerging, which we can call the religious cosmology.

We can diagram these three cosmologies as follows. The solid arrows imply a perceived direct influence; the broken lines imply influences that might be voiced but that nobody believes really have any actuality.

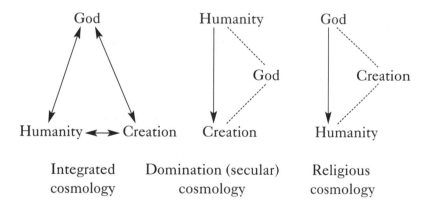

| Integrated cosmology | Domination (secular) cosmology | Religious cosmology |

The first model represents a cosmology that is largely absent from modern-day thought. The pervasive influence of science has left little room, particularly in the common culture, for this way of thinking. However, there are signs of a rebirth of this integrated cosmology in books that are being written and read, including this book.

The domination cosmology represents the prevailing modern

Western humanistic, secular culture. The integrated cosmology of St. Francis has been rotated so that humanity is now on the top. God is off to the side safely in Heaven and connected by dotted lines to humanity and the creation. Humanity, from its perch of understanding and domination, is in an exploitive relationship to the creation.

The other inheritor of the split following the Plague, the Western religious world, would diagram its understanding differently from that of the secular world. In this cosmology, God is safely back on top. Now God and humanity are linked to the creation by a dotted line. Religion has very little to say about the relationship of humans or God to the creation. The focus of most religious thinking in our day is on the relationship of humans to each other and humans to God. This is true of both progressive and conservative thinkers. Just check the titles at a local bookstore.

Implicit in both the domination and religious cosmologies is a separation of humanity from the creation. This reflects a dualism rooted in Bacon, Descartes, and Calvin, and still influenced by the original shock of the Black Death.

I was sitting in a class entitled the History of Christian Thought during my second year of seminary. Someone had asked the professor how God was related to the creation. I sat there as a budding theologian and leader of religious institutions. But, more significantly on this occasion, I sat there as a scientist. I had earned an engineering degree and had for four tough years learned to think scientifically. I had studied chemistry and nuclear physics. I had been interested in astronomy. I had a pretty good idea of how the world worked, *as described by science*. It was a complete system of thought— a cosmology, consistent in its methods and conclusions. It came straight from Bacon, Descartes, and their successors.

I had long ago passed the stage where there was a conflict between science and religion. The religious system of thought, which spoke of sin and grace, the relationships between people and between humans and God, was also a complete system able to speak to every new idea that came into it. I had a growing understanding of, and excitement about, that system.

The problem was that the two systems rarely connected with one another. Mentally, I had to make sometimes elaborate translations to hook them together. In my understanding the two systems were not in conflict; they were disconnected, like two languages from different linguistic traditions. One could interpret back and forth, but the connections were not readily apparent. So when the subject of creation came up in class that day, I leaned forward and listened intently. This could be the link between my two worlds. The professor talked about God creating the world. He talked about the creation coming out of nothing and the theological implications of that. He talked about the fact that everything created is good and the implications of that. And then he finished.

I raised my hand with a question. "How, then, is God involved in the natural processes of life that are described by chemistry and physics?"

The professor thought and finally said, "God created those processes, but that's about all we can say."

The class was over, and I was left pondering the gulf between science and religion. My mind was full of questions. Doesn't God have anything to do with natural processes after the beginning creation? Is God just concerned about humans? Is God willing to intervene at the human level but not at the level of the natural creation? Are humans that much different from the natural world? Where does the human spirit fit in? Can science talk about the spirit any easier than religion can talk about chemical reactions or laws of physics?

THESE QUESTIONS GET us to the crux of the matter. The issue is about wholeness. *Universe* means the whole creation, the whole world. It is one and unified. Science should talk about the whole. So should religion. To split it in two using an intellectual dualism is to violate the whole creation. It is to distort the way we think enough to lead us to damage ourselves and the creation.

Now, we must be careful. Actually, the world is full of opposites that express many dualisms: black-white, light-dark, before-after, known-unknown, masculine-feminine, suffering-joy. Possibility comes from the tension between these opposites. The problem comes when one of the opposites dominates the other and robs the tension of possibility.

This one-sided dualism is the problem. In our attempts to analyze and understand a whole, unified system, we often describe things using their opposites. But then we give one part domination over the other, squelching the possibilities that come from the creative tensions between equal parts. Destructiveness comes from giving one part greater value than the other. We have split up human beings into physical and emotional/mental parts, with different medical models describing each part. Modern medicine is still fighting about which half is primary—whether the body controls the mind or whether the mind controls the body. Research points to incredible connections between these two parts, but we still have different specialists with different systems treating the different parts and fighting over their value. This is one-sided dualism.

We insist on making a value separation between good and evil, physical and spiritual, mind and body, when there is actually one, unified system. It is easy to see the influence of Descartes, with his separation of mind and body ending in the primacy of mind. That one-sided dualism controls our modern thinking from the vantage point of 400 years ago.

If our minds allow us to split up the creation, making the physical part of it inferior to the spiritual or mind part, we can then exploit that inferior, physical part. Because of the influence of modern science, we humans see ourselves as outside of nature, with nature as the thing to be tamed, nature even as the enemy. Because of our modern religious heritage, we see God as concerned only about humanity and in the context of Heaven rather than the Earth. Religion in the past few hundred years has had very little to say about the rest of creation except as it fights rearguard actions to oppose new scientific ideas such as evolution.

The separation of science and religion by a dualistic cosmology where neither has to bother with the other has created the "Endangered." Before we knew it, our thinking was not just an intellectual curiosity, but a present danger to our very existence. Our way of thinking was giving us unknowing permission to destroy the unity of the creation, putting the creation and our own species in grave danger.

Recalling the stunning question at the end of the extinction chart in San Diego, we ask ourselves again, When humanity?

A New Cosmology

Let us now look at the seeds of a new cosmology that might heal the split. It started for me when I read *The Phenomenon of Man*[4] by Pierre Teilhard de Chardin (1881–1955). In the mid-1960s, I was vacationing on the Outer Banks of North Carolina, in Nags Head. I had heard of Teilhard in seminary but had found no time to get to him until that summer.

Teilhard, a Jesuit priest and a paleontologist, wrote in the 1930s and 1940s. He got into trouble with the Church in his latter years and was banned from publication until after his death. He studied fossils as well as theology. He devised a way of looking at the universe that embraced both my scientific side and my theological side, so that they united as long-lost lovers rather than warily circling as suspicious antagonists. Teilhard gave me hope of being intellectually whole instead of split with a fatal one-sided dualism.

In this seminal book, Teilhard agrees with science that the universe is composed of an almost infinite number of atomic pieces, each with its own particularity. But he sees that each piece has an influence, no matter how negligible, throughout the whole universe so that the basic reality of the universe is that of a complex system. At its heart, this whole complex system is held together by an energy that transforms the system toward greater and greater complexity.

Teilhard really riveted my attention when he suggested a better way of looking at the creation than just looking at its "without," as scientists do. They objectively measure, weigh, and categorize. Teilhard pointed out that the matter in the universe also has a "within,"

---

[4]New York: Harper Torchbooks, Harper and Row, Publishers, 1959.

a subjectivity, a consciousness. This "within" is most evident, of course, in human beings.

Teilhard says that if consciousness is evident in humanity, then it must be present in some degree in all of reality, from the beginning to the present and continuing on into the future. If there is unity in the universe, then consciousness couldn't be an exclusive property of humankind. It is present—at least in potential, in anticipation—in all of reality, from the simplest atom to the most complex molecule to the most complex system of molecules, the human. Just as an elementary school student can memorize the multiplication tables, so a newborn and even a fetus can memorize those same tables, at least potentially. We'll examine this much more closely later.

Teilhard goes on to point out that there is a discernible progression in the universe from simplicity of organizational structure at the creation to a rich complexity of structure in the present universe, not only in humans but also in many other forms. That same progression is accompanied by a growing consciousness from the beginning to the present. The external measure of increasing complexity is a growing internal consciousness.

Teilhard finally projected all of this evolutionary movement toward a point at the end of time, which he called the omega point. It sounded to me on that Nags Head beach suspiciously like Heaven.

So here was an intellectual theory that combined science and religion, the material and the spiritual, into one scheme that could comprehend everything. I was stunned and excited. But the press of other business kept further exploration on a side shelf until that day at the San Diego Zoo.

At that later time, sensitized by my shock at the zoo, I learned of Thomas Berry, one of the most articulate advocates for the place of humanity *within* the Earth system. Berry served as the president of the American Teilhard Society for the Human Future, and as such continued the work of Teilhard into the present.

Berry asserts that there are three principles upon which the universe has functioned from its beginning up to the present: differentiation, subjectivity, and communion. The principle of differentiation says that, from the beginning, the universe has not been an amorphous smudge but has been composed of distinct, differentiated radiation and particles that have evolved into much more complex elements, molecules, and organisms, into specific and differentiated stars and galaxies, and, finally, into this unique planet Earth and its unique life forms, including humanity in all of its particularity.

The principle of subjectivity points to the interiority of all particularities. It is the "within" of Teilhard. Berry, like Teilhard, points to the fact that subjectivity is increasing as the differentiations become more complex. With the human being, subjectivity takes on self-determination as in no other place that we know of in the universe.

The third principle, communion, points to the relatedness of each reality in the universe to every other reality. This is one big system. It is a unity and not just a random collection of unities. Each differentiated subjectivity is just a part of a larger whole, which is, itself, a differentiated subjectivity.

THE FOUR DYNAMICS

If we are going to heal the various dualities that have brought us the environmental crisis—the split between science and religion, the split between mind and body, the split between objectivity and subjectivity—it is important to have a common language with common categories. If the contending points of view can start with common categories, then there is hope for unity. If, within my own thinking, I can use categories that give equal access to both science and religion, then there is a chance for healing.

All of this is background for the introduction of the four *Dynamics* as the basic outline for a new cosmology, a new way of thinking. The Dynamics are heavily influenced by Teilhard and Berry, but were chosen to connote movement and for clarity and ease of understanding. They were chosen, fundamentally, to show the essential relationship of God in Christ to the whole emergence of the universe. They thus provide a way to understand the world from the points of view of science and the Christian religion simultaneously. To the extent that Christianity reveals the true nature of reality, the four Dynamics are the outline for a cosmology that reflects a unified way to perceive reality. Of course, it would be possible for the same thing to be done by any other religion.

The first Dynamic is the *Being Dynamic*. Every creature, every rock, every gamma ray has its own being. It is unique, created to be what it is in all its particularity. This rock is not like any other rock; and rocks, in general, are entirely different from seashells. Each being has limits, content, and discernibility.

Furthermore, each being is not just objective but has an interiority, a subjectivity. This subjectivity connotes freedom at some level. In the somewhat dated model of the atom, where electrons circle a nucleus, even the simplest atom with just a few circling electrons has a degree of freedom, as the electrons seem to jump from orbit to orbit without cause. They exhibit a certain elementary freedom.

As the complexity of beings increases, so does the interior subjectivity, until consciousness and even self-consciousness are seen in higher mammals such as humankind.

So the Being Dynamic points to the universe as composed of a collection of autonomous entities, each with uniqueness, each with not only an exterior but also an interior.

The second Dynamic, the *Belonging Dynamic*, points to the interconnectedness of all beings. Beings are never found all alone, but are arranged in systems that operate as wholes. We will call them belongings. Each belonging is composed of many beings and is, in fact, a being in its own right.

There are molecules, organisms, and organs. There are solar systems and galaxies and clusters of galaxies. All beings are composed of numbers of smaller belongings.

Furthermore, as the belongings increase in size, they increase in diversity. There are not nearly as many types of atoms as there are types of mammals. Solar systems are of necessity less complex and, therefore, less diverse than galaxies.

Being and belonging are always in tension. No being is ever as autonomous as the name might imply. Each being has universal influence, however minute, so that the distinction between a being and its belonging is quite blurred. Each of us humans is, on the one hand, an autonomous person with all that is involved in coming to terms with that. At the same time, we live in a society with all its demands, pushes, and pulls.

A simple diagram where the arrow denotes relationship might illustrate this:

Being

Belonging

The third Dynamic is the *Becoming Dynamic*. As implied earlier, all beings and belongings are in process. They are becoming. There is change in the universe. That change proceeds in a particular direction, as we shall see. The third Dynamic gives direction, movement, and growth to the universe, making it dynamic, not static. More than simply dynamic, it is purposeful. It is going from one place to another with intent.

In fact, we shall see that the universe is becoming more and more complex. The beings and belongings of the later universe are quite a bit more complex than those at the beginning. This third Becoming Dynamic reflects a universe headed toward increasing complexity.

The universe is also becoming in the direction of subjectivity or consciousness. An elemental consciousness resided in the earliest beings at the creation. That consciousness has grown with increased complexity to the self-consciousness evident in humankind. So the universe is not only becoming more diverse (see above), it is also becoming more self-conscious.

So we have being, belonging, and becoming. Finally, there is the *Surprise Dynamic*. Surprise is at the heart of becoming, describing the shape and thrust of becoming. Time and time again, there is a crisis, a falling away of one way that particular beings and belongings are organized. It is a death of sorts. Always, the universe reorganizes itself in a new, more sophisticated, more elegant manner. That new way comes as surprise.

There is, thus, an element of the unexpected in this unfolding. Out of the end of one life system, new forms will arise in ways that no one could have predicted. Fancy and exotic systems and beings emerge. So it often is with our own lives. When we think that all is lost, new avenues open in utterly unpredictable ways, bespeaking the universal drive toward life in the midst of death. The discerning Christian will note where we are headed.

Surprise is no surprise to the emerging universe. It is only surprise to humankind. Although surprise could be submerged within becoming from the universe point of view, it is awarded status as a separate Dynamic because it is so crucial to the human enterprise.

So that is it—being, belonging, becoming, surprise. Those four Dynamics describe the unfolding of the universe. We might diagram them as follows:

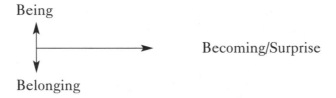

These four Dynamics are building a new cosmology that describes a universe that includes both science and religion. This new cosmology contains both objective and subjective reality, is dynamic and always unfolding, gives humanity a comprehensible place within the larger system, and begs for both scientific and theological treatment.

A Place for Theology

The task of theology has always been to explicate the human experience of God in terms of the prevailing cosmology. The human culture always has a cosmology. We have already discussed our prevailing Western cosmology and the damage it is doing to us and our environment. It is bringing us to the point of being endangered.

This book will suggest a new cosmology using the four Dynamics. Can theology translate our experience of God into the terms of this new cosmology? How would we understand Christ? How would the life and action of Jesus be seen through the lens of this new cosmology? How would we understand the biblical record of the dealings of God and humanity? How would we describe the church and the sacraments? What would faith development in a believer look like?

All of these and more questions will come pouring out, tumbling over one another as we believers search for new ways to believe in a universe that we are always perceiving in a new way. It is not that the universe is changing that much, although it is, of course. What is changing faster is our understanding of the universe. The ecological crisis is dramatically accelerating that change in understanding. It is almost faster than we can tolerate.

Our response to the environmental crisis must be rooted in our way of thinking, in our faith, if we would have any possibility of making an adequate response. This is probably true of the whole of humanity. It is surely true for those rooted in the Christian faith. Without our faith as part of our response, we are less than whole; the steam is taken out of us; we are crippled at the very place we begin.

So, this book is an attempt by one person of faith to push the darkness back a bit, at least for himself. What remains is to tell the story of the universe using the four Dynamics and opening the questions to which theology must speak. The effect, we would hope, will be to contribute to a discussion that will give some people within Western culture a unified sense of the purpose and direction of the universe and how we as a species and as individuals within that species are to live.

Chapter 2

The Big Bang Beginning

IT WAS AN EARLY spring trip to the mountain house of a friend. My wife, Carol, and I arrived and were delighted with the small cabin nestled in the woods on the side of a hill. We got ourselves moved in and the stove kindled against the chill night air. Outside, we could hear water trickling down the hillside as the latest of the winter rains drained out of the earth.

The next morning we went for a walk. The woods were still brown with winter; there were no leaves on the trees; there was only brown grass in the fields. Everything was somber.

Overhead, however, the sky was blue and the sun broke through the chill with energy for spring. As we walked, we found ourselves concentrating more and more on the little details around us. We saw nestled in the side of a hill little purple flowers just opening up. They were so delicate, so tentative. As we looked carefully at the ends of tree branches, there, as if swollen by juices from within, were small buds getting ready for life. Then we noticed that little green shoots were pushing aside clumps of earth here and there as if struggling to stand up and be counted. Ferns were thrusting up curled green shoots from their stalks so that it didn't take much imagination to see that they would open into proud fern fronds. Here and there a butterfly danced. Even some old moss was stirring with a new green at the tips of old growth. What we saw amidst the old brown of the winter was life. And it was everywhere.

Without realizing it, we had gotten there the first day of the spring opening. We had gotten there at the beginning. The landscape appeared brown. But on examination, it was seething with life, bursting, thrusting, opening, stretching to make itself known. The whole countryside was alive with an emerging extravagance that we could not have orchestrated or even dreamed up. This beginning left us breathless.

~&

THIRTEEN TO 15 BILLION years ago, in an explosion of unimaginable immensity, the universe came into being. Where there had been nothing, there was now something. Where there had been the void, there was now possibility. Where there had been brown, there was now green. What was it like at that first beginning? How did this universe start? Modern science is giving us a tantalizing view of that moment and the ensuing aeons. This compelling story engages us in mystery and possibility.

AT THE BEGINNING

At the beginning, the universe was pure energy, hot and compact. Computer models that extrapolate present observations back into the past can lead us right back almost to the beginning. They calculate that at a trillionth of a trillionth of the first second the density was 10^{50} grams per cubic centimeter and the temperature was 10^{20} degrees Centigrade. But with the incredible outrushing of energy, after one second the temperature had dropped by half and was only 10,000 million degrees Centigrade. This is about 1000 times the temperature of the sun but the temperature that is sometimes reached in a hydrogen bomb explosion.

It is almost impossible to conceive of this beginning. Theoretically, we are being asked to think of the entire universe compressed into a ball that you could hold in your hand. But you couldn't actually hold it in your hand, for its weight would be of almost infinite magnitude. Of course, there wouldn't be a separate hand for the holding either.

More personally, what we can say is that everything that was ever to be in the universe was present there at the beginning. Nothing new has been created since, so what we have now is what we had then. That means that we can look at any part of ourselves—say, our thumbnail—and know that it was there at the beginning. To be sure, it was not thumbnail. It was just thumbnail in potential. But the stuff that has become thumbnail was there at the beginning.

The content of this early fireball was pure energy such as light, gamma rays, and X-rays and elementary particles such as protons, neutrons, photons, mesons, and their corresponding antiparticles. These early elementary particles were torn out of a deep well of potentiality, particle and corresponding antiparticle, in an incredible cornucopia of abundance. However, in that swirling fireball, when a particle met its matched antiparticle, it was instantly annihilated into an explosion of light, only to be immediately followed by the birth of yet another pair of particles. This fierce birthing and consuming continued as the infant universe expanded and cooled, until the background level of light was no longer able to precipitate new particles. All of this happened by the end of the first second.

In the next two seconds, the universe foreshadowed its future course. It got more complex. Protons and neutrons attracted each other, bonded, and formed the nuclei of the light elements—helium, lithium, and deuterium. But before any larger nuclei such as carbon, oxygen, iron, and so on could be formed, the expanding universe blasted them apart and continued as an outrushing fireball of these light nuclei, photons, and various other elementary particles. The formation of light nuclei was an absolutely essential movement in the evolution of the universe.

For 300,000 years, the universe continued in this manner—expanding, cooling, gathering the conditions for the next great leap. By this time, temperatures had decreased to a few thousand degrees Centigrade. Then another remarkable and historic change occurred that was a surprise at that time and has never been repeated. These light nuclei attracted electrons to themselves like north and south pole magnets. The negative electrons positioned themselves in force fields with the positive nuclei into the first atoms. The combined size of these atoms was 10,000 times bigger than either particle had been before. Before, when the fireball was hotter, if the two had tried this kind of dance, they would have been immediately blasted back apart. But in the relative coolness that had come, they could stay together. The universe became more complex and positioned itself for the many evolutionary delights to come.

Several years ago I attended a conference at which Dr. Brian Swimme, Director of the Center for the Story of the Universe at the California Institute of Integral Studies, San Fransisco, was speaking. I'll never forget that morning when he told the story of the first atoms and then opened to us the "subjective" side of that surprise event. It planted the germ of the idea for this book.

Even though the universe had no central nervous system, role-play yourself inside the event as it was happening. The infant universe (less than a half million years old) must have been getting "anxious" that its energy was being dispersed so widely. What if it could no longer regenerate itself with the annihilation of those elementary particles? What if the universe were dying? We're speaking poetically, of course. But it would seem a possibility that the whole affair would just die out. It would run down with no new possibilities.

At that moment of crisis, a whole new thing happened. Rather than a continuing maze of charged particles whizzing around, dominated by raw energy, neutralized atoms now emerged as the dominant motif. Surprise! The universe had previously been dominated by raw energy. With this surprise, the dominant reality would quickly grow to be matter. With this transition from dominant energy into dominant matter, the building blocks for all of the future were in place. With this transition, the possibility of humanity was created. With this change, the business of building a cosmos could now proceed.

THE SURPRISE DYNAMIC

The universe is full of surprises like the creation of atoms. Looking back at that or any other surprise event, we sometimes wonder why we hadn't expected it. But on the early side of the event, we don't have a clue. Just like the disciples struggling over the news of the resurrection, we struggle with the newness that is continually erupting in the universe.

The newness seems always to come out of the death of the old. Just when we get used to things, it seems, they die. There is anxiety.

What will we do now? Again and again the universe brings new life, new forms, new ways of doing things. Its not just that the new supplants the old, like fall and spring on the Earth follow one another. It is that, of course. But, more important, in the universe something new is repeatedly invented that is like nothing before. The invention of the new always happens in the context of the ending of something else. It is as if life and death flow in and out of one another, jumping on and over one another, in a continual climb of surprise.

Albert Einstein often said that the ultimate question in the universe was whether or not it was benign. Was it good? Or not? Surprise is the affirmation that it is good, that the universe is pushing toward life, and that life invariably comes out of death.

A story circulated recently about a young girl who was made a refugee by the war in Vietnam. All she had known was her native land and the war. One winter she was flown to Memphis, Tennessee, so she could be safe and lead a normal life. But when she got off the airplane, she panicked. Looking around, all she saw in the wintry landscape were denuded trees. She thought that the war had come to Memphis and Agent Orange had been dropped on the city, defoliating all the trees. The fall browning and the spring surprise of regenerating green were unknown to her. The young girl was quickly reassured about the trees and told about spring. Her hosts had forgotten to warn her, having taken the surprise of spring for granted.

Finally, the truth about surprise is that we don't really understand it until it is worked personally in our lives. We can understand it intellectually as a concept. But we only really get it when it happens to us in our experience. The young Vietnamese girl would forever after understand surprise.

I can still remember with some pain the time my marriage came apart. My wife had moved out. My children were confused. I was in a deep funk, desolate, at the end of my rope.

A friend said to me, "Why don't you be good to yourself? Why don't you go see a psychiatrist? I went to one once and he literally saved my life." He gave me a name and a phone number.

The thought of getting help had occurred to me, but I was terrified. What if the psychiatrist confirmed my worst suspicions that my life was just a sham and an illusion? What if he saw me as a hopeless failure? What if he agreed that the weaknesses I suspected I had were terminal weaknesses? Could I stand that knowledge on top of the rejection in my marriage?

The phone call to the doctor was, perhaps, the hardest thing I had ever done. To my dismay and relief, he had an opening and I was to see him the next week.

Terror hung at the fringes of my mind during the ensuing wait. Finally, the day arrived. On the drive to his office I had an image that he would take off the top of my head, peer in, and it would be full of maggots.

After we had met for a number of weeks, a couple of things started to occur to me. First, I was feeling better. But second and more important, I did have some maggots crawling around in me but, what the heck. There were maggots everywhere, and I was OK in spite of the maggots. The stunning news was that in the face of my dread, life—my life—was wonderful. Surprise!

Surprise! The universe does bring life. It is good. That I learned, deep in my bones.

❧

The Becoming Dynamic

BUT THIS UNIVERSE story is surely more than one of just surprise. There is also an unfolding in it that reflects the Becoming Dynamic. This universe is alive, changing by surprise. And it is directional, moving toward complexity and diversity, toward consciousness and life. The direction is also purposeful. All of this is the Becoming Dynamic at work.

The very first expression of the Becoming Dynamic was in the creation of the universe. There the reality began. According to science as well as theology, at one moment there was nothing and then there was something.

It was inconceivably dramatic, but what else could we expect? It started quite simply but then moved quickly toward complexity. It defies description—an incredible beginning density, unbelievable heat, an expanding explosion that would make anything we humans even dreamed of seem like a popgun. In the land of America where we celebrate records and bigness, this one dwarfs them all. Physically, it was beyond belief.

Beyond its physical dimension, perhaps the more interesting dimension is the view that comes from looking back through the event into the mind and heart of the Creator that was driving the creation. This Big Bang must be seen as sheer giving, profligate generosity, a pouring forth that would strike terror in the heart of any of us for fear, if we were to try it, we would end in absolute depletion. It was a generous overflowing, a cornucopia of extravagance, a heart brim-full of love that just had to give itself expression.

Perhaps the nearest thing to this creative energy is the work of great artists. George Frederick Handel, for instance, wrote the entire *Messiah* in 23 days, overflowing with creative inspiration that kept him going night and day. Other artists report that the creative energy in them must find expression. So it seems to be with the Creator of all that is. Deep in God's being, generated out of profound compassion, is the drive to become.

THE QUESTION FOR THEOLOGY

As Christians, we assert that God was behind the creation of the universe. God was the generator of the becoming, the force behind the surprise. But who is this God that does this creating?

One thing we must make plain. This God we talk about is not static; this God is not cast in stone. This God is in process, moving, always leaning forward. This God is dynamic.

We are used to a very static image of God. We are used to seeing God as strong and unchanging. That static image is rooted in the Middle Ages. In that day, cosmology portrayed an orderly universe ruled over by an orderly, omnipotent God. The Earth was at the center, the sun, moon, and stars revolving around it in stately procession.

Heaven was up. God was on a throne in Heaven, ruling like a benevolent emperor over life on Earth. God even had a long, flowing, white beard. Just ask any child.

Thomas Aquinas (1225–1274), the great theologian, gave theological expression to that cosmology. Aquinas said that theology was based on revelation and reason. God was the only perfection and was the adequate cause for the order found in the universe. God was described as omnipotent, omniscient, all-good, immutable, and eternal.

The feudal system, with its regimented, hierarchical place for everyone, kept society functioning with a peaceful harmony. Gothic architecture provided vaulting arches and soaring interiors to give an inspiring lift to the whole order. A visit to Chartres Cathedral in France confirms the medieval synthesis that sees that God was up in Heaven and all was right with the world. Chartres, in its splendor, is even a glimpse of that Heaven.

The science, the theology, the economy, and the art of that world all conspired to say the same thing: everything had its place in an orderly, unchanging universe. Deep in our cultural memory we assume that this is the truth and the only truth about God and the world.

Of course, it is the truth. God is eternal, the Unmoved Mover. But as with all great truths, there is a paradox involved. The whole truth has another side. This same Unmoved Mover God is also the restless God of the people of Israel, the people on the move across the Red Sea and through the desert. This God is also eager, searching, striving, vigorous, and dynamic.

Look at the world of today as contrasted to the Middle Ages. Science is anything but static, as we shall be seeing throughout this book. Modern science is often based on uncertainty, not certainty. Einstein with his theory of relativity and Heisenberg with his uncertainty principle are in; Newton, with his strict cause-and-effect universe, is out.

Our free enterprise economy is built on change, movement, growth. The art of our day is always new and experimental. Going to an art gallery, to a concert, or to the theater, one never knows what to expect. It has been said somewhat wistfully that the only constant in our day is change.

Many people living in this shifting modern world look for a God who is solid, stable, and unmoving. They want absolute, unchanging answers in a world of flux. But it also makes sense in this changing world to explore an understanding of a God who is dynamic, moving, changing. Only that understanding will give ultimate coherence to the world of change in which we live. We have not changed God by so doing. God will be God no matter what we think. We have just looked at a side of God's character that will let us understand our world in a new and sustainable way.

The question for theology is to describe this God and the activities of this God in a way that is dynamic and moving. What would we, who know that the universe is becoming, say about God? Could we describe a God who authors a becoming universe that is punctuated with surprise? Is it possible to root the Surprise Dynamic and the Becoming Dynamic in the very essence of God? Can a case be made that God is dynamic as well as static? Can that be said in words and images that are understandable to the culture of our day? To do this is one of the foremost challenges to theology.

But there is more to the question of God. There is also the question of the "place" of God. This new expanding universe stretches out in all directions. How can we describe the place of God in a dynamic universe like this? Where do we "see" God? Is it up? Is God higher than we are? The former premier of the Soviet Union, Nikita Khruschev, stated that the Soviets had sent spaceships into outer space and were able to report definitely that there is no God. We counter quickly that we speak of a spiritual place when we speak of Heaven. But it still gnaws at the back of our minds that "up" is where outer space is, not where God is. So where is God?

Paul Tillich, a twentieth-century theologian, had an image for the place of God that might be useful. Tillich imaged God as being at the depths, at the core, at the profundity of life. He described God as "the Ground of Being."[5] For Tillich, God was not being but the foundation upon which all being rests. Without that foundation, the creation would collapse as a building would fall down should its foundation be yanked out. For Tillich, God was "down."

---

[5]*Systematic Theology*, vol. 1 (Chicago: University of Chicago Press, 1951), p. 156 and others.

Now, of course, when Tillich said "down," he didn't mean literally down. Neither did Aquinas mean literally up. But spatial direction in our imagery of God can mean the difference of credibility as we acknowledge and worship the God that is.

Perhaps there are other, more cogent images of the activity and place of God. The feminist theologian, Sallie McFague, in her book, *Models of God*,[6] speaks of the Earth as the "body of God." She is speaking metaphorically, of course. But she is suggesting that it might help us to see God as "permanently present," as close to the world as we are to our own bodies. God is more than the world, of course, just as we are more than our own bodies. But this image suggests an identification that would give new meaning to loving God. It would give new imagery to the "place of God."

What contributions might other theologians, who look at things from the female point of view, have? Process theologians, who view the world in terms of change and process, also have much to contribute. Liberation theology looks at the world from the point of view of freedom for oppressed people. Think of the contributions they could make to this discussion if they applied their ideas to oppressed species.

It is crucial that we craft an image of God that will overcome the dangerous dualism we talked about in the first chapter. God must not be removed from life. God must not be outside of life. God must be involved in and at the very root of all life. At the same time we are sealing this immanent, involved nature of God, we must be sure that we do not lose God's transcendent nature. It is a tricky balance. But it is the key to a new cosmology. It is the key to the survival of humankind.

In the beginning God, in a mighty act of compassion and self-completion, created the heavens and the Earth ...

GALAXIES AND STARS

Now, let's get back to the exploding universe. After the surprise formation of hydrogen, the universe continued its rush outward. Then

---

[6]Philadelphia: Fortress Press, 1988.

another surprise occurred. The outrushing matter, hydrogen and some small parts of helium, started to group together by gravitational attraction in areas of the universe that were slightly more dense than others. In those areas, a gravitationally induced collapse started and, due to other gravitational pull from outside those regions, each of these clusters started a slow rotation. As the collapse continued, this rotation increased until, all over the universe, gigantic galaxies were created. We think there are about a trillion of these majestic whirling spirals of stars and stardust. They were all created at about this time—500 million years after the initial explosion. None seems to have been created since.

The universe could have kept up its outward rush without ever congealing into any formations that would have brought a substantial base for life. What had been working for a billion years could have kept on going. But a new thing happened that had never been before—galaxies! Now that is astounding. That is a universe surprise. Who could have dreamed it up?

It happened so spontaneously and so universally, it reminds me of a true story. A man went fishing in a small mountain lake out west one winter. As he quietly walked down the path to the lake, the temperature hung at about freezing. It was early in the morning, and not a breath of air disturbed the cold and tranquil scene. Upon arriving at the lake, the fisherman cast his line out over the water. As the hook hit the water, the lake instantly froze from shore to shore in a thin but solid sheet of ice. Surprise!

It seems that the water had been supercooled by the cold air but needed a slight disturbance to tilt it into freezing. The hook provided that disturbance, to the fisherman's astonishment. So it might have been with the early universe. Conditions were just right for the consolidation into galaxies. Slight irregularities throughout the whole system became the occasions for the organization of that matter into galaxies. Surprise!

More was happening. Within the swirling galaxies, hydrogen and helium atoms were breaking up into smaller clouds, collapsing together and heating up in the process. As they drew together, nuclear furnaces were ignited so that the force of the inner nuclear explosion balanced the force of the gravitational pull. Thus were created stars.

The fires in these stars came from the thermonuclear burning of hydrogen. The by-products were heat, energy, helium, and the heavier elements, such as carbon, iron, and uranium. Those star furnaces were the great incubators for more sophisticated life to come, sucking hydrogen and helium into their burners and generating the building blocks of heavier elements that would be needed later.

After a few billion years of burning, the stars would collapse, heat up in the collapse, and then explode again in what is called a supernova. These explosions set in motion a regrouping of matter into new stars and, sometimes, planets orbiting those stars. Our sun is a second- or third-generation star that has produced enough heavier elements to sustain life on at least one of its planets, Earth.

SCIENTIFIC THEORY OF RELATIVITY

Before we go any further, let's look at the two theories of physics that have led us to these conclusions about the origin of the universe. These theories, the theory of relativity and the theory of quantum mechanics, drove the rest of twentieth-century science. Prior to these two theories, Newtonian physics explained everything in the physical universe. The formulation of the theory of relativity provided a much better explanation of the movements of giant bodies such as galaxies and the behavior of light and gravity over giant distances. Quantum mechanics explained things on the very minute subatomic particle level, along with the behavior of light and energy at that atomic level. Newtonian physics is still very useful for everything in between. Even today, science has no unified theory that will explain it all, but people like Stephen Hawking are working on it.[7]

First, let's look at relativity. This theory says several things. Fundamentally, it says that the universe is expanding in all directions. It is getting bigger and bigger, as if part of a gigantic explosion. But it is more than an explosion, because that connotes a slowdown as time progresses. The universe's expansion is unabated. The universe is incredibly dynamic. In the 1920s, Edwin Hubble and others exper-

---

[7]*A Brief History of Time* (New York: Bantam Books, 1988).

imentally demonstrated this fact by measuring the light waves coming to us from distant galaxies. The wavelengths of these waves were greater than Newtonian physics might have predicted. This discovery proved that everything in the universe was moving away from us. The universe was expanding!

By knowing the rate and direction of expansion we could, therefore, project backward in time to a point at which the expansion started. At that beginning point, 15 to 20 billion years ago, the universe was of infinite density and zero size. The explosion from that point was later called the Big Bang by the British astronomer, George Gamow.

Before the twentieth century, it was broadly assumed that the universe was static. It was neither growing nor shrinking in size. It was assumed to be the same eternally, world without end. Albert Einstein, the German physicist, in developing his theories of relativity between 1905 and 1915, worked largely by intuition. Those intuitive hunches led him to postulate an expanding universe. But he couldn't believe it. He, along with everybody else, believed that the universe was static. So he inserted a fudge factor into his mathematical formulas, which allowed for a static universe! It wasn't until a few years later that another mathematician pointed out that the fudge factor was unnecessary and that we are indeed living in an expanding universe. Hubble then experimentally verified this.

Not only is the universe expanding from that original Big Bang, but it is expanding equally in all directions from all points. As we look out with huge telescopes, we see that everything is moving away from us at the same speed. We are at the middle of the universe! But lest we fall back into that medieval trap which said that the Earth is the center of everything, relativity leads us to the assertion that everywhere in the universe is the center of the universe! Not only is our galaxy the center, every galaxy is the center. We are in an omnicentric universe!

Our minds start bending out of shape at this point. It is hard to contain or imagine what is being said. One way to start to get a handle on it is to think of a balloon being inflated. Every point on the balloon is moving away from every other point as the balloon is being blown up. So it is with the expanding universe. Mathematicians do it by talking of space-time warps and other things bordering on science fiction. All most of us can do is accept what they are saying and marvel at the fact that science and religion might be finding a place again at which to speak a common language.

Remarkably, the universe is expanding at just the right rate for life, as we know it, to emerge. If it expanded much faster, it would go off into outer space so quickly that gravity would not have been able to pull enough back together to form galaxies, stars, and worlds. If it had been one part in one hundred thousand million, million slower in the rate of expansion, the universe would have recollapsed before reaching its present size. We humans wouldn't have had time to emerge. At any other rate of expansion, something else would have happened. It just wouldn't be this world with its present life.

Two other conclusions from the theory of relativity must be mentioned. First, there was no preexisting space or time. Space and time are, therefore, relative and are unfurling with the expansion of the universe. It is easy to understand that time would be generated with the expanding universe. It is not much more difficult to imagine space growing between exploding galaxies. We could even imagine space growing between expanding molecules. But then try to imagine "before" the Big Bang. There is no time, or at least only "negative time." Whatever the limits on our imaginations, the universe is not just static. It is quite dynamic. It is an emerging event.

We must also point out that, with a relativistic universe, there is no objective point of view. Surveyors can establish benchmarks on the ground that precisely measure latitude, longitude, and the height of the mark above sea level. These then become standards from which to measure the position of other objects.

In this new understanding of the universe, we cannot find a center and measure everything else in relation to that. There is no edge either. There is no absolute place in space. Everything is relative. Or to say it another way, we need every point of view in order to measure or understand any one thing.

Now a brief theological word. From the point of view of the Becoming Dynamic, things start to get real interesting. The universe is becoming. Since it is omnicentric, no matter where we look into the universe, we are looking into its center—namely, into the heart of God. God is at the center, and God is everywhere.

Since the universe is expanding at precisely the right rate for life, as we know it, could we not say that God is expressing God's self in a way wholly consistent with the determination that life would emerge? From the point of view of life, therefore, God has absolute integrity and purpose.

QUANTUM MECHANICS

Before we look at the second important twentieth-century scientific theory, quantum mechanics, it is important to establish the scientific context for it: Newtonian mechanics. Isaac Newton (1642–1727), an English mathematician, invented modern calculus and did important work on light theory and gravitational theory. He is best known, however, for his laws of motion, which state that objects always move in a straight line unless directly changed by another force, in which case they always react in an equal and opposite direction. This theory established the mechanical cause-and-effect world that has dominated the Western—if not the whole—world for most of the past three centuries. Newtonian mechanics, which explains so much, also has caused us to set up a rigidly predictable world with cause and effect, action and reaction, in every possible corner of life.

Quantum mechanics is changing all of that. Here's how. The German scientist, Max Planck (1858–1947), suggested in 1900 that light energy was not emitted from matter continuously, but in small packets called quanta. This theory finally explained the confusing observable data on the emission of radiation from hot bodies that had scientists of that day so puzzled.

In the 1920s, another German scientist, Werner Heisenberg (1901–1976), devised his famous uncertainty principle and then applied it to mechanics, formulating a new theory called quantum mechanics. It applies to very small, subatomic particles, but has revolutionized all of science.

Quantum mechanics states that it is impossible to measure the exact position and speed of any particle. The most we can do is give a probability of the position and speed. If we can't do it for particles, which are the building blocks for all matter, can we predict the position and speed of anything, including planets and galaxies? NASA's space program counts on us predicting with great certainty the future position of bodies in space. But Einstein's work showed that we needed more than a mechanical explanation for objects at high speeds. Putting it all together, we must say that the universe is not strictly mechanical. It does not operate with rigid laws of cause and effect, as Newton had postulated. All we can do is predict the probabilities of events. The universe is filled with uncertainty!

The established way of looking at light, for instance, was to see it as a wave that bounced rigidly and mechanically off a flat surface back at the observer so that he or she could see an image of the object. Further exploration of the quantum phenomenon, however, shows that energy does not reflect off an object like a mirror seems to reflect light. Rather, it is absorbed by the object and then reemitted, in quantum amounts, just a fraction of a second later. Instead of just reflecting, therefore, it can be said that every object gives off waves of energy—proportionate to what it has absorbed, of course. Every object is shining like a light bulb, not just reflecting like a mirror. Every object is pulsating, expressive, revealing, exhibiting itself.

When we talk about energy, in fact, we now sometimes find it easier to speak of energy as quantumlike particles. At other times, energy behaves more like waves, such as light or heat. At the very small atomic level of things, the distinction between energy and matter blurs. This equivalence was stated most famously in Einstein's well-known formula, $E = MC^2$, where E is energy, M is mass, and C is the speed of light.

Now let's look at the implications of this in relation to the Being Dynamic, which states that each being is unique in terms of not only its external form but also its own interior consciousness. We can define consciousness as the ability of any being to be aware of and respond to its surroundings. Therefore, according to quantum mechanics, when we look at any being in the universe, we are not just seeing a reflection of light or heat or some other energy. We are, in reality, seeing an expression of that being itself into the space beyond itself. This means that the universe is filled with self-expression, pointing to some degree of consciousness in every being in the universe. The universe is not just cold and objective. The universe throbs with inner subjectivity, consciousness. We will explore the concept of consciousness more thoroughly in Chapter 4.

It is possible to ground this consciousness in the very life of God, a task that theology can tackle. In the beginning, in a mighty act of self-expression, consciousness, God created the heavens and the Earth. For the first time in several hundred years, modern theology and physics would be starting from the same place as they told the story of the universe. It is now possible for science and religion to develop the same language and be reunited as they had been before Copernicus and Galileo.

Before moving on, we must briefly note one further area of scientific inquiry. During the past few decades a growing number of scientists from widely scattered disciplines such as meteorology, biology, ecology, physics, and mathematics have been developing a discipline called chaos theory. Their contention is that normal, natural processes are never predictable by Newtonian cause and effect. Rather, there is motion like we see in clouds and running water. They conclude that all natural processes are marked by turbulence, not predictability. Even the rhythmic swing of a pendulum, Exhibit A of Newtonian mechanics, has a limited turbulence. They say that exact predictability is nothing more than a figment of the imagination in the minds of humans.

It is evident that the becoming universe is not just a mechanical process of cause and effect. It is alive. It has freedom and possibility. The universe is at least an organic progression of more or less probable events. It is also a majestic, compelling, beautiful, unfolding display of the Creator God.

THE SHAPE OF THE UNIVERSE

This universe is 30 billion light years across. That is easy to say because, if it started 15 billion years ago, then the news of that happening is spreading outward at the speed of light and has been doing so for 15 billion years. Light travels at 186,000 miles per second. So by simple computation we figure it would travel 175,970,880,000,000,000,000,000 miles in 15 billion years. That is the radius of the universe.

The alert reader will now point out that we have just said in our discussion about relativity that there was no edge to the universe. So how could we measure how far it is from one side to the other?

The reader is right. Since every point is at the center, the universe curves back on itself. Look at it this way: we think of a three-dimensional globe as having no edge and no center on the surface; you can sail forever and never fall off, and you're always at the center of the surface.

Think of the universe as having four dimensions: time and the three dimensions of space—height, length, and breadth. Imagine the universe as a four-dimensional globe. There is no edge to it, and everywhere is the center. Except that it is 30 billion light years across. (I just couldn't resist that last paradox!)

There are about one trillion galaxies in that universe. In fact, most of these galaxies are clumped together in a number of groups. The Virgo cluster, of which our galaxy is a member, has about 1200 galaxies in it. Our galaxy is associated most directly in a twin relationship within the Virgo cluster, with the Andromeda Galaxy, which is a mere two million light years away.

Our galaxy is called the Milky Way and has a minor star within it, which is our sun. Each galaxy is shaped like a saucer, and spins around a highly concentrated mass of matter at the center that is dense enough to be a black hole. A black hole is so dense that its gravitational force will let nothing escape, not even light. It just sucks matter and energy into itself.

When my dad died, he left a little money so my wife and I and four of our teenagers took a Caribbean cruise. I'll never forget the first night out as the ship sailed across the heart of the Caribbean, far from any land. Despite the bright lights on the ship, we could see stars in abundance overhead. Carol and I decided to walk up to the dimly lit bow of the ship for a better look. There the stars literally jumped out at us. The heavens were alive. Our breaths were taken away at the majesty of the sight. Amidst the display, a dense band of stars spread out from horizon to horizon. We were looking at the Milky Way, our own galaxy, from edge to edge. It is a sight rarely seen by city dwellers anymore because of polluted air and the reflection of city lights. In the middle of the Caribbean, the universe is alive and vibrant.

꙳

THE MILKY WAY galaxy has about 100 billion stars in it. It is 100,000 light years across and about 15,000 to 20,000 light years thick. The sun is about 30,000 light years from the center, so we probably won't fall into the black hole any time soon. Our solar system revolves around the center of the Milky Way about once every 200 million years. That means that during the entire period of time that human-like creatures have been on the Earth, five million years, the solar system has gotten only one-fortieth of the way around the galaxy. Does this leave you a little anxious, wondering what might be out there waiting for us on the backside of the galaxy?

BEING AND BELONGING

The miracle of all of this, of course, is how it holds together. It is an example of the Being and Belonging Dynamics. You'll remember that being speaks of the authenticity and uniqueness of each separate being. And belonging speaks of the membership of each being in a larger system, just as each system is in itself a being and at the same time a member of an even larger system.

Each part of the universe has its own being and role. It is what it is. It can be and should be no other. Each part has integrity. We might even say that each part is called to be what it is. If it doesn't play its appointed role, then the whole works will wobble out of balance.

But the Being Dynamic also relates to the integrity of movement of beings. Let me explain. Despite Heisenberg's uncertainty principle, Newtonian physics still gives the best approximation of the movement of all beings above the atomic level and below the galaxy level. In Newtonian physics, we say that an object has inertia. If it is moving, it will stay moving in the same direction and at the same speed unless another object or force, such as friction, gravity, or a brick wall, interferes. Then its inertia will change. Its speed and direction will change. This is the Being Dynamic in operation. Because of inertia, or the Being Dynamic, any being has a built-in commitment to keep moving in the direction, and with the speed, it has been moving.

The Belonging Dynamic means that every single being in the universe is a part of a larger being. Molecules belong to rocks, which belong to continents, which belong to planets, which belong to solar systems, which belong to galaxies, and so on. The larger being operates as a system of smaller beings, each being faithful to its particular calling. If one or more of the smaller beings doesn't perform, then the larger system is affected and will have to go through the pain of adjustment. This is also true of a human being. If our liver fails, then the whole body is affected. Perhaps the body can adjust. If it can't, then the whole will fail because of the failure of one of the parts to be what it was supposed to be.

Belonging is expressed in Newtonian physics through the principle of attraction. The beings in the universe attract other beings, through electrical force, gravitational force, or magnetic force. This attraction is pervasive and holds beings together in their larger belongings. At the same time each being must follow its individual calling. It must do what it is meant to do.

Being and belonging are, therefore, always in balance and tension with each other, as illustrated by twirling a rock on a string around in circles over your head. Being wants the stone to continue in a straight line out into the next-door neighbor's yard. Belonging pulls the rock in toward your hand through the string. They are inextricably linked and in perfect balance. Planets stay in orbit as a balance between the Belonging Dynamic, which pulls them together by gravitational attraction, and the Being Dynamic, which propels the planet out into space through its inertia, or centrifugal force. Galaxies also swirl in stately fashion throughout the expanse of the universe, as examples of the Being and Belonging Dynamics at work.

The Solar System

About five billion years ago, out of the debris from a previous exploding supernova, the sun and its planets were born. The sun's nearest
star neighbor is Alpha Centauri, which is four light years away. With
the means we presently have at our disposal, it would take us 100,000
years to get there, which is longer than the entire history of *Homo
sapiens.* We can't count on escaping any degradation of our planet by
going to the nearest star and hoping for a hospitable planet there.

The solar system is about 7.3 billion miles, or 11 light hours, across.
The sun is a medium-sized star that has a mass that is 333,000 times
the mass of the Earth. The Earth is about 93 million miles or eight
light minutes from the sun, going around the sun every 365.2 days.

The Earth is midsized in relation to the other planets of the solar
system. The four inner planets—Mercury, Venus, Earth, and Mars—
are smaller and are Earth-like, with landmasses, present or previous
volcanic activity, and so on. The four giant planets past Mars are
largely frozen or metallic states of what would be gases at earthen
temperatures. The largest of these, Jupiter, is one-tenth the size of
the sun, and thus, 33,000 times bigger than the Earth. Even when
they are on the same side of the sun, the distance from Earth to
Jupiter is four times the distance from Earth to the sun. That is why
it appears in the night sky as no more than a bright star.

It is possible to construct an imaginary model to illustrate the scale
of the solar system. If we think of a model with the sun being at
Times Square in New York City, and the other planets arrayed out
over the continental United States, the position in the model and the
relative sizes are as described in the chart on the following page.

One thing that impresses even the most hard-bitten of us when
we talk about these kinds of things is the enormity of it all. These
are small bodies, suspended in vast distances. Despite the enormity,
an incredible transfer of energy and exertion of influence goes on
between the bodies. In fact, for human life to be possible, the solar
system couldn't be much smaller. If it were any smaller, there would
be no place for humans. If it were any smaller, butterflies couldn't
fly and flowers couldn't bloom. If it were any smaller, life would not
be possible because the odds would increase greatly that there

Planet	Position in the Model	Size in the Model
Sun	Times Square, New York City	extends out to Hudson River
Mercury	Morristown, NJ	a standard automobile
Venus	15 miles from Pennsylvania	a two-story house
Earth	seven miles into Pennsylvania	a two-story house
human beings		a thin coat of paint on the house
the moon	three football fields from Earth	a VW bug
Mars	Reading, PA	a large sports utility vehicle
Jupiter	Pittsburgh, PA	a 30-story building
Saturn	Chicago, IL	a 25-story building
Uranus	in a cornfield, 100 miles west of Omaha, NE	a barn
Neptune	Salt Lake City, UT	an urban residential lot
Pluto	San Francisco, CA	a small piano

would have been a collision of heavenly bodies before this moment. We need this much room in order to live.

Perhaps the most awe is evoked from us by the sheer extravagance of it all. The Earth gets about one-billionth of the sun's energy. The rest goes into outer space and is "wasted." Yet that one-billionth is enough. Flowers bloom in utter profusion, despite what we do or don't do. Sunsets irradiate the sky, and fish stir the deep. Worms crawl through the earth, and sequoias reach into the heavens at a scale that is breathtaking. The universe is nothing in the unfolding if not extravagant.

That takes me back to the trip to the mountains when the whole Earth was coming alive with a spring extravagance. Everywhere I could imagine, and then in places that I couldn't imagine, life was pushing up and out into shape and form. It was an extravagant becoming, celebrating the creativity of God, reflecting the 15-billion-year emergence of the becoming universe.

THE EARTH

Between four-and-a-half and five billion years ago, out of the dying of an unnamed older star in an exploding supernova, the solar system with the infant Earth was formed. The ball of matter that gathered to form the Earth was hot, heated by nuclear radiation generated in the supernova explosion and regenerated by the crush of the weight of the matter piled up on the Earth's molten core. But the cold atmosphere of space cooled the outside of the ball, allowing a hard crust to form. Inside it remained hot, and that heat exploded through the crust in volcanoes of lava and hot gases. The lava hardened to form a thicker crust, heaping up in places to form huge mountains. The gases rose to start a primitive atmosphere. Some of those gases cooled into water vapor, and the cooling rains began. Other gases stayed in the atmosphere. But no human could have lived there long because, until life came, there was little oxygen.

Hundreds of millions of years went by. The inner core cooled a bit; the crust thickened; gases arose and the rains fell. Water rushing down those early mountains caused the first erosion and gathered in low places to form the first seas. Evaporation from the seas began, and the rains started to wash minerals into the seas, gathering the ingredients needed for life.

The stage was set for the next becoming. Organic life, which had been present only as potential, was about to burst forth as an extravagance that could have hardly been dreamed of before. Out of the dying of a star and the cooling of the debris, the surprise of emerging life was about to unfold.

CHAPTER 3

Explosion of Life

HAVE YOU EVER SEEN the surprise of new life? All of us have seen the first blush of spring, the birth of kittens, even the hatching of a baby bird. But let me tell you of the surprise of new life as I experienced it . . .

It was happening in the backyard. A young mother-to-be was lying on a blanket under a tree. Her husband was holding her from behind. The two older children of the family were excitedly looking on. A young midwife was calmly orchestrating the whole affair. And I was there, taking pictures!

Believe it or not, since I had a hobby of photography, I had been invited to take pictures at the home birth of the child of a young family in my parish. When they asked me to do that, I had been skeptical. How would I know that I would not be busy when the moment of delivery came? Would that be the kind of thing I would want to be in on, anyway? Wasn't this too private an affair? And—how can I say it?—would my sexuality stay under control when I was openly faced with the most private sexual parts of a strange woman? Those concerns do cross one's mind. But after much discussion and thought, I agreed to be there if I could.

When the call came, I was free, so I grabbed my camera and drove out to their home. Everybody was in the backyard, although they had originally planned to have the birth in the bedroom. It seems that they started the early labor in the back yard to stay cool, but then things progressed so fast that they were unable to move inside.

I set about my task of picture taking, relieved to see that, despite the position of the woman, this was not a sexual event at all. It was more of an earth event.

I hadn't been there long when the pace of things picked up. We were coming to the moment of birth. Contractions came closer and closer to one another. The measurements of dilation came at shorter intervals. Suddenly, the top of the head of the baby, with a little snatch of hair, could be seen. What a wonder!

Before we knew it, the head emerged, then one arm, and then the other. And then the whole baby. A little boy, umbilical cord still tying him to his mother, covered with the debris of nine months in the womb and . . . blue.

It was only a few seconds, however, before there was a little gasp, and the whole baby changed in a wink to . . . pink! I was speechless. A lump caught in my throat, and my knees got weak. It was, perhaps, the most incredible thing I had ever witnessed. Life flushed into that child as he switched from umbilical cord to lungs. All we did was watch and wonder, awestruck at the flush, the miracle of new life.

SOMEWHERE BETWEEN three-and-a-half and four billion years ago, the Earth flushed. What had been a soup of cooling chemicals, atoms, and primitive molecules finally developed the complexity and sophistication to reproduce parts of itself in the form of tiny protein molecules. It was a new life system. How could that have happened?

In the years preceding, in order to form the building blocks for this self-reproductive life, the Earth drew energy from the sun and its own internal fire, concentrating that energy in lightning and lava. It circulated the necessary ingredients, using the process we all learned about in grade school science. There was first the constant evaporation of water into clouds. Then the water vapor in the clouds condensed into rain, which collected in streams and seas. Through that activity, needed chemicals washed from the mountains into warm incubator-like sea shallows, where they collected for the event of life.

Meanwhile, the Earth moved masses of material around, with the grinding and sliding of continental plates and the eruption of new matter from its bowels, through volcanoes, geysers, and springs. Meteors from outer space smashed in through the thin atmosphere, adding new matter to the infant Earth.

In the gathering soup of those primitive seas, energized by lightning, lava, and sun heat, chemical reactions between atoms and molecules combined these different elements into ever-more-complicated molecules made up of carbon, hydrogen, oxygen, nitrogen, chlorine, sulfur, and a few other elements. Through the millennia, these combined molecules grew in size and complexity to form amino acids, sugars, proteins, and DNA and RNA molecules.

Finally, in a surprise invention that leads directly to the reality of our own existence, the DNA molecules, with the aid of free enzymes, split themselves in half. Each half then attracted smaller molecules to fill in for the half that had left. Lo and behold! Where there had been one molecule, there were now two identical molecules. The drama of self-reproducing life had begun. The Earth flushed into biological life.

It is like going back to a high school chemistry lab. Do you remember those little round blobs of mercury that the teacher tried to keep us from losing? Each blob was perfectly round and independent. But if you split one in half with a pencil each of the halves would be perfectly round, just like the original had been. There was always the class wise guy who would splatter the blob with his finger. Resulting were many, many little round blobs, each a perfect, albeit smaller, copy of the original. Of course, the analogy breaks down because the mercury blobs can't do it by themselves. It takes the class wise guy.

Perhaps a more poetic way can describe the Earth coming to life. In those early seas, the vitality of the God of the becoming universe thrust up through the primal elements toward an expression of the creation that could dance. That dance would expand, multiply, and develop into a mighty ballet in later years. That dance could evoke only the gasp of wonder. Like life pushing up through a blooming flower, God pushed up through the creation in an ever-growing wonder that found life begetting life. That life had heritage and continuity and future.

LIFE'S STORY

We must be under no illusion that this new ability of molecules to split into like molecules in that primal sea marked the beginning of life. It did not. From the Big Bang beginning, God had been pushing life up through the expanding universe. Life was simple there at the first—just raw, untamed energy. It was nothing more than energy or light waves and elemental particles.

Then there was hydrogen and helium, representing an immense jump in complexity, where particles paired and lived in an embrace that was tough and durable. Then there were more complex atoms and then the combination of atoms into molecules. Then gravity pulled these molecules into more complex systems, like stars and planets and galaxies. Finally, on at least one planet around one star, the conditions were right for the complexity to grow to a state where molecules could reproduce themselves. This was one more invention in a long series of inventions, each a surprise, driving toward what we call life.

We so want to separate life from nonlife, animate matter from inanimate matter, and make life and animate matter superior. That is a dangerous dualism, as dangerous as placing mind over body, or spirit over matter. There is no superiority, despite the domination that Descartes sought to portray. Let me illustrate.

A nun was hauling stones to the top of a hill to form a rock garden on her farm. Cows mooed in the distance. Ripening corn rustled in the soft breeze. It was summer and very hot. The stones were heavy.

The nun struggled until she thought of the deeper meanings of her task. Rainwater would leach minerals from the rocks at the top of the hill. Through the years, those minerals would be carried down the hill by the rain. There, the minerals would collect in the pasture and be absorbed by the grass. The cows would eat the grass and make milk. That milk would be fed to babies to make strong bones. So the nun was not just hauling rocks. She realized, with a start, that she was hauling babies' bones. The rocks suddenly got lighter under the hot sun.

Babies' bones are made from the stuff of rocks plus some protein. The material of babies' bones has been configured in a way to allow the organisms to reproduce themselves. We call it life. It might more properly be called organic or biological life.

There is directionality in the evolution of the universe to biological life. It is the direction of growing complexity. From gamma rays to self-reproductive biological life in the early sea, we can observe an immense journey toward complexity. From our perspective, we know that this early sea life is only the beginning. It is like meeting people. At first, they might seem to be simple and one-dimensional. But as we spend time with them, we discover depths upon depths of complexity. So it is with the becoming universe: the longer this story goes on, the more complex becomes the expression of life.

Growing complexity brings with it growing consciousness. From the rudimentary freedom in the energy particles of the early universe, to the dull consciousness of the first biological life in the sea, to the more developed consciousness of humankind—all is a steady progression toward complexity and consciousness. The more complex an organism; the more possibility of consciousness in that organism.

My men's book group was meeting in the living room of my house. We had been gathering about once a month for several years. We took turns hosting the group and the host chose the book we would all read and discuss. My choice was the manuscript of this book. I wanted some fresh eyes to read it. One of the members was a biologist and another an anthropologist. I would value their reactions and corrections.

Generally, they liked the book. But then the conversation turned to consciousness. We agreed that a workable definition of *consciousness* was "the ability to be aware of and react in some way to our surroundings." My assertion that some form of consciousness was present even in the fundamental particles in the earliest universe drew a rather heated discussion. The group said that consciousness could not be present before biological life existed and that it was important to establish that point.

"But didn't we agree that *consciousness* could be defined as the ability to be 'aware' and then react?" I replied.

"Yes."

"And weren't those earliest energy particles in the universe doing just that?"

"Yes."

"Then that must be called a primitive form of consciousness."

The group was not easily convinced. Nor am I. Upon reflecting on the conversation the next day, it occurred to me that this is the very crux of the Western dilemma. We insist on giving a higher priority to biological life and insist on separating it from inorganic matter like stones and energy particles. But it is actually one continuum. It is all life, albeit at vastly different levels of complexity and sophistication. It was an example of the one-sided dualism of Western thinking.

My book group, without knowing it, redoubled my determination to proceed with this project.

ॐ

A Progression of Forms

BUT LET US get back to the story. Those early self-reproducing molecules were known as microbes. They were very vulnerable, particularly to ultraviolet radiation from the sun. So they soon learned to form protective walls around themselves, which could shield them from the radiation. Those walls also had the ability to take in new raw materials and eliminate wastes. These walled microbes were the first one-celled creatures, and are still the most numerous form of organic life on Earth. They are called bacteria.

Actually, over the first two or three billion years of organic life on Earth, three general kinds of bacteria developed. They became the basis for all of ensuing life. The earliest bacteria, developed some four billion years ago, used a process of fermentation to grow. These bacteria, the so-called bubblers, lived in warm, shallow seas and fed on plentiful supplies of sugar and acid molecules floating around in those seas. They gave off, as waste products, gases such as methane and carbon dioxide. The making of bread and wine relies upon this same earliest bacterial process, as does the production of bubbles in mud and natural waters. It may be no coincidence that the elements for the Holy Eucharist—bread and wine—embody the most primitive life-giving processes of the Earth.

But over time, the supplies of food in the sea became scarce because of the proliferation of these bubblers. They were eating food faster than it could be produced. Without the necessary food supplies, how would they obtain energy? How would they keep from starving to death? A crisis had developed.

The becoming universe had an answer. There was carbon for food all around in an atmosphere saturated with carbon dioxide. So some of those early bacteria "invented" a way to use the energy from sunlight to break down the carbon dioxide and combine the parts with water and the plentiful rocksalts in the sea to produce the necessary food sugars and DNA parts. These new bacteria might be called "bluegreens"; the process is photosynthesis; and the inheritors are, of course, the entire kingdom of plant life. That process has gone largely unchanged for over three billion years and still uses carbon dioxide, water, and minerals to produce abundant green life all over the planet. So out of the food shortage crisis came the second type of bacteria—bluegreens.

But as is so often true, the solution to one problem brings another problem. Over a period of two billion years, the waste gas from that new type of bacteria accumulated to a degree that again imperiled all of life on Earth. It turned out that the waste gas from plant life, when overly concentrated, was a deadly poison that could burn up even the giant molecules of the living things that produced it. It was more deadly than anything earlier, even ultraviolet radiation. The new deadly poison was oxygen. Again there was a global crisis. How could the amounts of oxygen be controlled to acceptable levels?

Again the becoming universe had an answer. Yet a third kind of bacteria, the so-called breathers, was developed. These bacteria could, through a new process, use the accumulated oxygen in the air to break up food molecules and turn them into protein and energy so that they could live and grow. The process was called respiration. All animals, including ourselves, have descended from these new bacteria. The remarkable added benefit is that the waste product from this process—carbon dioxide—is the very staple of the bluegreens, the photosynthesizing plant life. Out of the crisis came a wonderful cooperative system, first among the primitive bacteria swimming throughout the seas and then in the complex animal and plant life that swarmed the whole planet. We will talk more about that later.

But please notice that in the becoming of the universe, new forms were produced that could hardly have been predicted. Time and again, in the midst of a crisis, a new surprise emerged. Even from our perspective of hindsight, we can do little else but marvel at the way the story emerged—new forms, new processes, new complexity. Always new. Out of a crisis, always surprise.

But we have hardly begun the story. Remember that new bacteria were still being generated by the split into two of existing one-celled bacteria. Parent bacteria were splitting and becoming their own children. Over time, the process built on itself with an increasing growth we call exponential growth.

Natural death had not yet been invented, so none of these bacteria just naturally ended their lives. They might starve to death, get eaten, or be killed by radiation or a poisonous gas such as oxygen. But they didn't die because they had lived their "three score and ten." So when a system got worked out in which they were helping each other like those three types of bacteria, natural growth could be quite amazing and rapid.

If you have ever played the old game of choosing whether you would take $10,000 or the results of starting with one penny and doubling it each day for a month, you know the consequence of this kind of growth. You'd better choose the penny, for by the end of the month your penny doubled 30 times would be worth $10,737,422.24.

You can start to get a picture of the results of this kind of growth, even in the vast seas of the Earth. It went on at exponential growth rates for two billion years. By then, the planet was crawling with one-celled bacteria and it became hard to keep up a food supply for this kind of population explosion. Competition for food became quite intense. Not surprisingly, some of the bacteria started attacking other bacteria, burrowing inside them and eating their internal food stores.

The invaded bacteria naturally fought back. But as often as not, the invader won, ate up the host bacteria, and then perished from lack of food because it had killed its own food source. It is a story we humans know all too well.

Out of that crisis came a new surprise. Some of the hostile bacteria started finding ways to cooperate. They took the best features of the two competing bacteria and combined them to form a complex one-celled organism that was much more versatile and, therefore, more stable. That is another theme throughout the creation. Complexity produces stability.

Over time, other kinds of bacteria found their ways into the increasingly complex organisms. Some performed the function of energy production; others produced food. Some developed primitive methods of travel, while still others provided defenses against hostile neighbors. The gene codes of life that were the DNA in the early one-celled creatures combined to form complex chromosomes. Over time, these chromosomes became the nuclei of these complex new cells. Each nucleus served as the coordinator for its whole organism. Eventually, multicelled organisms evolved with an incredible ability to perform many functions.

The miracle that allowed the multicelled organisms to develop was cooperation. Instead of fighting to the death, the becoming universe demonstrated that cooperation was the key to survival. This was the Belonging Dynamic at work. All of us learn this central truth, in one way or another.

I learned it when I became the rector of a parish that had really been struggling. There had been membership loss, income loss, and, most important, the loss of spirit. In my second or third year, the parish finally had a good pledge drive and had more money than it needed for its minimum expenses. The fight that developed between competing interests over the extra money was classic. Some wanted a new building. Others wanted to start a ministry to the poor. Still others insisted that we get a new organ. Each item was crucial to the life of the parish, and each was expensive.

I can remember saying throughout the showdown meeting that if we were to do any of them, we would have to do them all at the same time. It might take longer to get it all done, but by cooperating all of the tasks would eventually get done. The alternative was to dilute the energy of the parish through a competition where the winner took all—an alternative that would probably split the parish.

In the end, we agreed to do all the work at the same time. The truly remarkable thing was that the total package was accomplished in about three years. If we had done them one at a time, it would have taken at least 10 years.

IN ADDITION TO all of the competition that goes on, life in the parish, as well as life in the universe, is a cooperative business. Each unique part is a member of a larger system. The destiny of each part is absolutely linked with the destiny of the whole system. There is no survival alone. And the secret is not only the cooperation, but the diversity. With a diverse parish all working together it took three years rather than 10. Cooperation is a law of the universe.

But, let us get back to the story of the continuing emergence of life. The cooperative life in the sea was developing an incredible complexity. While single-celled creatures were still the majority, as they are even today, there was more and more experimentation with multicelled life. Mostly the new creatures were small, multicelled algae that got around by crawling or wriggling or flapping. They were soft like plankton or jellyfish.

But with the development of multicelled creatures came another important advancement: the invention of death. As we noted earlier, single-celled organisms never die a natural death. But on a planet with limited resources, something has to give. Any progression of forms calls for old forms to die out.

So in the emerging genetic structures of organic life came genetically produced old age and, finally, death. Thus, creatures could die, go back into the earth, and become food for succeeding creatures. So was born the first recycling project.

It was not until about 540 to 550 million years ago that organic sea life developed some hard parts for itself. The earliest of those frames were external shells. Later those shells were internalized in the form of skeletons. The hard shell made it possible for the first time to leave a good fossil record in the rocks.

Fossils occur when an organism dies and sinks to the bottom of the sea. There, it is covered by silt and other dead creatures. Over time, considerable layers of matter stack up. The pressure of those layers plus seawater then compress the lower layers of sediment into rock. Many millions of years later, convulsions of the Earth's surface can bring the rock above sea level. When the rock is split apart, the impression of the early creature is revealed as a fossil. Paleontologists can then date the rocks using carbon 14 or uranium half-life dating techniques and reconstruct the story of the creatures that lived those millions of years before.

The same fossilization process can occur on land, of course. There the dead creature often sinks to the bottom of a mud slough and is covered with layers of mud and sediment through the years. Again, as the layered rock reemerges at a later time, it can be "read" for the history of its fossilized inhabitants.

So we see developing, through the reading of the historical fossil records, an intricate and balanced biological life system on the Earth. Breeding, eating, evolving, dying—life goes on. Flowing, evaporating, blowing, condensing—water carries the stuff of life across the face of the Earth. Decaying, leaching, combining, growing—the material of life reprocesses into other living systems. It is all one big, interconnected life system.

GAIA

Among scientists, a new idea is gaining credence. More and more are saying that the Earth as a whole can be best understood as a living organism. They call this organism Gaia. The name comes from the ancient Greek Earth goddess and denotes that the Earth is a living thing, rather than a planet with life on it. This Earth has many different processes like the many processes that go on even in a single-celled organism. Each is vital to the whole, and the elimination of any one could affect the whole.

Dr. James Lovelock has led the discussion on the Gaia concept[8] and has pointed out in some compelling ways how this all works. He uses a model called Daisyworld. In that model, a global change in temperature is moderated by the type of daisies that grow. Hot climatic patterns encourage white daisies, which in turn reflect more sunlight back off the Earth and thus lead to cooling. Conversely, cool weather encourages black daisies, which absorb the sun's heat and promote a warmer climate. The Earth is like a house, and the daisies like the house thermostat. They keep the climate within a certain range.

Lovelock also points out that there is no way to separate living from nonliving matter on the Earth. He notes that much, if not virtually all, of the rock on the Earth has been at one time or another a living creature, deposited on the sea floor and compressed into rock. That rock is then leached out into dissolved minerals, which can easily become part of another life form. The elements of the air might just recently have been part of a tree, or flowing through some creature's lungs. Rocks, rather than being part of the nonliving Earth, are actually the skeleton upon which life can thrive. It is very much like a giant redwood tree. Only about one percent of the tree—the bark and the needles—is "living." The rest is "dead." But no one would say that the tree is not living.

There is, again, no adequate way to separate the living from the nonliving. We are part of one complex, interdependent life system. The Earth is Gaia, a living organism that flushed into self-reproducing life when those early microbes learned to split and multiply.

BACK TO THE DYNAMICS

It might be useful at this point to return to the Dynamics and to look at them again in relation to Gaia. If the planet is one system, then it is a belonging system. Everything is tied into everything else. An incredible balance of parts has been worked out over the aeons so

---

[8]*The Ages of Gaia* (New York: Bantam Books, 1988).

that it is a dynamic and relatively harmonious organism. The constituent parts of that organism are seas and atmosphere and rock, plant life, animal life, and insect life. Each of the beings that make up the whole is itself a system of belonging that is in turn composed of constituent parts. Therefore, we might say that beings are belongings of other beings, which are in turn belongings.

The human body works this way. The body is a being. But it is also a belonging, composed of organs and vessels, muscles and tissues. Each part of the body must operate as it was created to operate if the whole is to work. Toes must do the work of toes and ears the work of ears. The liver cannot say that it wants to do the work of the heart. If it tries to, there will be liver failure and the body will die.

Each part must, therefore, have a clear sense of being, reflecting the Being Dynamic. It must know its role. It must know who it is, what its purpose is. If a particular being gets too far out of line, then it can endanger the balance of the whole. Could that be what is happening with the human species in our time? Could it be that we don't understand our being or apprehend our belonging?

But the universe is more than a collection of beings and belongings. It is also a progression toward complexity. The progression in beings is toward increasing consciousness. In the simplest beings, consciousness is evidenced in the ability of each particle of matter to express itself in quantum emissions. As the story has unfolded, consciousness is more and more the ability to be aware. Beings learned to be awake to their surroundings. Then they developed the ability to choose. We even noted this phenomenon in subatomic particles, whose position and speed couldn't be pinned down. It was as if they were conscious and could choose.

Billions of years later, we see the ability to choose much more in animal life and most markedly in humans. We humans can be aware and choose. Despite our genes and social conditioning, we make choices. We are remarkably conscious.

But that is not the only story. Over the billions of years of the history of the Earth, an incredible diversity of beings has emerged. This diversity has gathered in belongings that are increasingly complex. So we say that there is a progression in belonging toward diversity.

The more diversity, the more stable is the whole system. If one of the beings should break down, the wide variety of other beings helps Gaia solve the breakdown in a number of ways. Diversity in and among belongings promotes balance and stability.

It's like a football team. If all the players are great blockers and no one can run, the team will not get far. A good team needs a variety of players. Some must run and some must block. Some must throw and some must catch. Some must be leaders and some must be followers. Some must be able to think creatively and some must just do what they are told. The balance of the interlocking parts will make a solid and stable team. So it is with the Earth. Great diversity of beings produces a solid, stable Gaia.

In summary, there are some discernible progressions in the story of the universe. There is a marked journey of becoming toward complexity. Beings are developing toward more consciousness. Belongings have been getting more diverse over the 15-billion-year story. This might be diagrammed as follows:

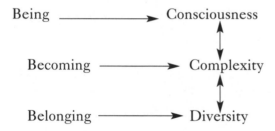

Increased consciousness is always in the context of, is always related to, increased complexity. It is as if a concentration of organized complexity develops an ability to be conscious. But consciousness is always a derivative of complexity, so that consciousness will never overwhelm being. There can be no disembodied consciousness, no spirit-over-matter dualism.

In the same manner, diversity also increases as complexity increases. Diversity is a derivative of belonging, but can never overwhelm belonging. Diversity is always a product of more complex belongings in order to stabilize those belongings.

All of this is the Becoming Dynamic at work.

EVOLUTION

Now it is time to describe our best understanding of the mechanisms of this becoming. How does God work in pushing the universe toward complexity in its consciousness and diversity?

We have to start with Charles Darwin (1809–1882). An Englishman, Darwin developed his theory of natural selection over his lifetime, starting with observations made while serving as the naturalist on a voyage of the *H.M.S. Beagle* from 1831 to 1836. Sailing around the southern tip of South America and particularly in the Galapagos Islands off the western coast of South America, Darwin observed the similarities and differences of related species on neighboring islands, as well as on the continent.

In Darwin's time, the prevailing orthodoxy was that species were created full-blown and unchangeable by God. What we saw was what God had designed and created. The idea of the extinction or evolution of a species was unthinkable.

Darwin changed all of that. First, he found some fossils of extinct animals whose skeletal structure was quite similar to that of existing animals. Then Darwin noticed that each island in the Galapagos had its distinctive species of bird, lizard, or tortoise. Why was God so active in creating different forms there in those remote islands?

Then Darwin noticed that the differences in species were closely related to the food supplies available. Where there were large seeds, the birds had large beaks. Where there were small seeds, the birds had small beaks. Where insects predominated, the birds had fine beaks. These observations planted the germ of the idea that species adapt to the conditions of the particular environment in which they live.

Full blown, Darwin's theory says that species evolve from earlier species through the process of natural selection. Members of a species will produce many and varied young. However, only those young most adapted to fit into an open, natural ecological niche will survive. Therefore, species always change to fit better into an available niche. Actually, the environment on this dynamic planet is always changing. Evolving species follow those changes. So there is constant pressure on the species to change to fit the new environments.

Darwin made no mention of members of the species competing with each other for a place in the environment. They were selected by the environmental conditions. That distortion of natural selection was created by a social scientist, Herbert Spencer (1820–1903). He coined the phrase, "survival of the fittest," and argued that evolution justified cutthroat competition for economic gain between individuals and classes.

Obviously, there is competition and conflict in nature. Individual creatures eat other creatures as part of the balance in an ecosystem. In mature, balanced ecosystems, eating and being eaten serve to keep the overall balance between species in that system. Natural selection means that adaptation, cooperation, and harmony are the overriding qualities that foster long-lived species. Those that fit in are those that survive. Those species that can best balance their being with belonging to the bigger system are those that survive. Unbridled competition has no place in Gaia.

Punctuated Equilibria

Darwinian theory has had no serious scientific challenges. Some paleontologists, however, have been puzzled about a couple of things that just didn't fit classic Darwinian theory. First, they have noticed that most species for which they have extensive fossil records change very little during the course of their existence. They will remain essentially the same throughout their whole history.

This gives rise to the second question. Given the wide variety of species and this slow rate of change, there just hasn't been time to develop the incredible variety and complexity of species that we have today. Could there be another explanation for how species come into being?

An explanation was given to me during my surprise at the San Diego Zoo. For there I bought a book by a paleontologist, Dr. Niles Eldredge.[9] Eldredge, in collaboration with Dr. Stephen Jay Gould, came up with a plausible answer. Eldredge specializes in trilobites,

---

[9]*Life Pulse* (New York: Facts On File Publications, 1987).

which are extinct, primitive relatives of horseshoe crabs and crustaceans. In the 1960s, Eldredge discovered a bed of sedimentary rock in upstate New York that contained the fossil record of a change from one trilobite species to another. It was a monumental find and led to the generally well-accepted theory of punctuated equilibria. That theory opens for us an understanding of how the Becoming Dynamic operates in evolution.

It works as follows. A discrete species will inhabit an ecological niche within a geographic range. There, that species can do the two things necessary for its survival. It can find energy for living, and it can procreate. As long as the environment remains the same, the species will remain the same.

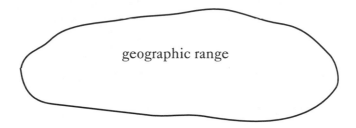

But on this changing Earth, environments seldom remain the same. There are warming or cooling trends. Continental plates shift. Landmasses rise or fall, are flooded or become dry.

Usually, a species will migrate when stressed, changing its geographic area to meet climate shifts, thus avoiding the necessity for major change. But often the environmental change is too abrupt, or migration paths are blocked. Then a part of a species can't migrate. Some members of the species are cut off from the main body. Those members will have to change or perish.

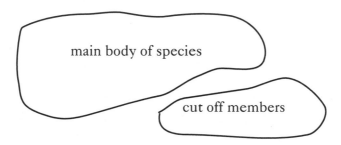

Often the separated part of the species will perish. Sometimes the main part of the species will perish, leaving an empty niche. During mass extinctions, wide environmental niches are left unfilled and these allow a burst of new experiments, called adoptive radiations, to fill those empty niches.

In the case of the trilobites, upstate New York was a shallow sea at one time. As a result of plate shifts, the European continent bumped up against North America and thrust up the Appalachian Mountains. Some of the shallow sea at the extremes of the trilobites' geographic range got separated from the rest of the sea. A few thousand trilobites were isolated in relatively harsh circumstances. They had to either change or perish.

Examining the fossil record showed that over the course of about 10,000 years, at the most, the trilobites moved from one species to another. At the bottom of the shale is the old type, exclusively. Throughout the middle, there are a wide variety of intermediate types, many kinds of experiments in different adaptations. At the top is the new species, exclusively. The new species developed in less than 10,000 years and then, firmly established, lasted for millions of years.

Eldredge called the theory "punctuated equilibria." Equilibrium is the norm for all species. They can stay unchanged for millions of years. But this is interrupted, or punctuated, from time to time with a swift change for small parts of existent species. New species are quickly formed. In a dynamic world, with change happening quickly in many places, it is enough to explain the variety of species.

The number of adapters who explore new ways to live in a new environment is small. The place where they do the new experiments is quite local and often undiscovered. So we have discovered very few examples of actual changes between species. In a world largely unexplored, this theory is enough to also explain so-called missing links. We just have very few examples of the changes.

The main actor in this whole scenario is the species. Species evolve, live, and then become extinct. True, a species is made up of many individual organisms. Those organisms must be similar enough to interbreed if the species would remain alive. At the time of environmental change, individuals within the species will live to the extent that they are suited to the new environment. Less suited individuals will perish. The main actors then are a few individual organisms within the species. When enough individuals suitable to the new environment are present, a new species is born. This is called speciation, and can take from 5000 to 50,000 years.

Once a few of those individual organisms make a suitable adjustment, they will multiply to form the new, sustainable species. That new species, in the absence of environmental pressure to change, can remain stable for 10 to 15 million years or longer. That harmony is the most common story of life on the Earth. The harmony of the species, of course, is also harmony for the organisms within the species.

We might construct a diagram to illustrate the theory of punctuated equilibria like this:

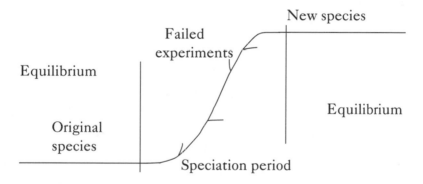

Not all paleontologists agree with Eldredge and Gould. Some point out that many species do not remain unchanged during the time of equilibrium but also adapt because of competition between individual members of the species. Favorable traits are strengthened. Unused traits tend to atrophy and disappear. An example is the human brain. According to fossil discoveries, the human brain has doubled in size over the past two million years. That can be attributed to the advantage that a larger brain gives to individuals within the species. In like manner, the appendix appears to be vanishing from the human species. It is no longer useful. The species is not new, but it is changed.

THE STORY OF LIFE (CONTINUED)

We left the story of the emergence of organic life about 550 million years ago (m.y.a.). In that time frame we have the first hard fossil record of creatures with shells. Life until then had been simple, with relatively few different forms. It had lasted about three billion years. Then about 540 to 550 m.y.a., some of those creatures at the bottom of the sea floor developed shells. Their presence created variety and opened up some important new ecological niches. Nature rushed in to fill those niches.

The first new creatures were largely trilobites. They dominated the sea floor for over 130 million years in many different forms and sizes.

Then crinoids evolved about 530 m.y.a. Like sea lilies, they are related to modern starfish. Corals came on the scene around 510 m.y.a., building their castles of rock on the shells of their predecessors. Around 490 m.y.a., ammonoids (an early relative of squid and octopus) appeared, as did the first fish with their distinction of having the first internal shell called a skeleton.

So in a bit more than 100 million years—a mere slip of time in the universal scheme of things—life in the sea was off and running with a growing extravagance. Different forms multiplied, fed and fed on each other, hid from and with each other, played on and with each other, and, in general, proclaimed the exuberance of the becoming universe.

Finally, some sea algae found their way up onto land and established themselves. About 425 m.y.a. a sea plant found a way to emerge out of the sea. It needed to develop an outer covering (or cuticle) so that it could retain water. It needed a vascular system so that water could go from its roots to its leaves and spores for reproduction. The Cooksonia did all these things and took the giant step to land, leaving a fossil record of the first sea plant moving to land.

A short time later, 395 m.y.a., some arthropods (early trilobites and modern insects) and some scorpions, too, became the first creatures on land. These early settlers were closely followed by fish, either lungfish or lobe-finned fish, using their fins to scurry across mud flats.

With all of these new environmental niches becoming available, life burst forth in a luxuriant variety of forms. Trees came about 370 m.y.a., followed closely by seeds as a form of propagation that largely replaced spores. Amphibians, which could live on land or water, appeared about the same time. The decay of trees and other vegetative matter produced the great coal swamps some 330 m.y.a.

Reptiles evolved about 313 m.y.a.; the dinosaurs, 235 m.y.a.; and the first mammals appeared as a minor form about 220 m.y.a. They weren't very important, however, because they were dominated by the dinosaurs until the extinction of the dinosaurs.

Some of the dinosaurs took to the air as the first birds some 150 m.y.a. Flowers bloomed into life about 110 m.y.a. and the first primates made their debut some 70 m.y.a. Vast grasslands developed in the neighborhood of 20 m.y.a. and the earliest hominid, the predecessor of *Homo sapiens*, emerged about 5 m.y.a. But that is a story unto itself and will be reserved for the following chapter.

EXTINCTIONS

This story would not have happened this quickly or unfolded this broadly if it hadn't been for species extinctions. Ninety-nine percent of the species that have lived on the Earth are now extinct. They are gone. Their biological codes, their unique DNAs, are lost forever.

Much of this extinction is what is called background extinction. Environments change in a small way but a species cannot change quickly enough. So it disappears. This is a regular event, occurring at about the rate of one species every 1000 years. The effect is usually to open an environmental niche to be filled by one or more new species. Life does not die with a species. It continues to emerge.

The dramatic story of extinction involves the great mass extinctions. There have been five major extinctions since the fossil record has been available. They have not only changed the face of life; they have made possible the blossoming of life.

The two biggest extinctions were during the Permian and the Cretaceous periods. The Permian extinction occurred about 245 m.y.a., when some 95 percent of the species on the face of the land

and in the sea perished. It was the end of that early pioneer, the trilo-bite. The crinoids were left hanging by a thread, represented by only a few species. Almost every species of coral disappeared. Fish were not too affected, but the ammonoids were cut down to a hardy few.

We know little of the cause of that extinction. Perhaps the onset of a period of glaciers lowered the sea level and reduced habitat areas. Perhaps low plate activity led to low mountain production and so to decreased runoff of life-carrying nutrients into the sea. Or per-haps a meteorite hit plunged the Earth into a severe climate shift to which few forms could respond. Whatever it was, environments were thrown out of whack. Ecosystems collapsed. Creatures that provided the food for other creatures perished, throwing the whole food chain into confusion. The disaster compounded and there was a mass extinction.

The other great extinction was during the Cretaceous period, about 65 m.y.a. It took out some 90 percent of the species, including the dinosaurs, which had been the dominant land animal for 145 million years. It also marked the end of the underwater ammonoids and a host of other marine species. Again, we don't know the cause of the extinc-tion, though evidence indicates a giant meteorite caused this one.

In addition to these two great extinctions, three other major ones are evident from the fossil record. The Ordovician extinction, about 438 m.y.a., hit marine invertebrates such as trilobites and early reef builders particularly hard. The Devonian extinction of 370 m.y.a. hit fish and invertebrates as it wiped out 70 percent of the Earth's species. Finally, the Triassic extinction, about 208 m.y.a., destroyed some 55 percent of the world's species. Life has not been easy on this shifting Earth.

What is most important to us is the result of these extinctions. Each extinction opens up many environmental niches so that many new forms can evolve. The Cretaceous extinction of 65 m.y.a. pro-vided the opportunity for mammals to emerge. This, in turn, paved the way for humans. Without that great death, we humans would probably not have had life. The Cretaceous extinction also resulted in a wild multiplication of fish species. There are now more species of fish than there are of mammals.

THE DYNAMICS AT WORK

What is going on in all of this activity of organic life on Earth? What is the meaning? Is it just chance, just a random sequence of events? Or is there direction and purpose?

There is a clear progression, as we have pointed out, from the free-flowing chaos and simplicity of the creation Big Bang to the growing order (still marked by chaos and turbulence) and complexity of organic life on Earth. The line will not stop with mammals, birds, and fish, but will push on through *Homo sapiens*. What is clear is that this is a story of the progressive becoming of the universe toward consciousness, complexity, and diversity.

Consider again that progression. At the beginning there was little sense of being in the creation fireball. It was frenetic, directionless. On the other hand, the consciousness manifest in a songbird or a deer or a whale has a clear sense of being. It is what it is and can be mistaken for no other. It does what it does and does it well.

At the beginning, the inchoate being exhibited in the fireball was balanced by an equally inchoate belonging. That early energy was there, but the relations of one part of it to other parts was not clear. It was swirling, unformed chaos. There was little order, little relationship, little sense of belonging to a larger system.

Later, the beings on the Earth were quite clear that they lived in a greater belonging. The songbird was dependent on insects and worms for food. Other life was dependent on the songbird. It was a part of a greater ecosystem of carefully balanced parts. Likewise, the deer and the whale fit within complex ecosystems. And those ecosystems were beings in themselves, fitting into a greater belonging called the Earth. That belonging is infinitely more complex and diverse than the beginning fireball.

We can also say that it is more and more clear that surprise is the driving dynamic of the whole progression. Again and again, one thing ceases and by surprise a new thing emerges. Species die and new, more complex species evolve. Ecosystems collapse and new ones come into being. By surprise, being moves in stages toward increasingly complex expressions. By surprise, belonging moves toward increasingly diverse expressions. By surprise, becoming is energized in the emerging of the universe.

Some would object that the law of entropy demands that the whole Earth system will eventually run down and die out. John F. Haught in his book, *Science and Religion,* points out that the breakdown of order, called entropy, is actually what allows new belongings to emerge with more complexity and diversity. He even calls it surprise![10]

❧

I learned how life comes by surprise from within a dying being through counseling a young woman. She had been abused as a young child, and the effects had left their marks. There was in her family, and later in her, a pattern of excessive drinking, promiscuity, and the inability to make a decision. When decisions were made, they were often self-destructive. She was marked for death.

I started talking to her weekly because, despite being in regular counseling, she needed all the support she could get. At times it was touch and go. But we prevailed, and she grew stronger and stronger.

The thing that most captivated my attention was the way she started making decisions. She would often ask me what she should do about a certain thing. I would ask many questions and often have a clear idea in my own mind of the course she should take.

But I had learned through years of mistakes not to give advice. People would often not follow it. And if they did and it didn't work, then they would blame me and have little sense of their own responsibility. I believed advice to be a bad thing generally, and I had grown to avoid giving it. But I usually left a meeting with this young woman with a clear sense in my mind of what she should do.

The next week, dying to know if she had done what I thought she should do, I would ask her how things had turned out. Invariably, she would have decided in some way other than what I had imagined. The surprise was that it was always a better solution than the one I had come to. Out of the little dying that comes with a decision would

---

[10]New York/Mahwah, NJ: Paulist Press, 1995, pp. 178–179.

emerge from her inner being an answer that was unexpected, unthought of, a real surprise. It was always life for her.

⚬

A THRUST OF LIFE up through the creation pushes toward greater and more complex life. It is inexorable. It is unbelievable. We can trust in it. But it will come by surprise. There is no predicting it and surely no controlling it.

A DIGRESSION ON CREATIONISM

Some religious thinkers in our day would vehemently disagree with most of the above. Creationism is a piece of Christian thinking that holds literally to the biblical word and insists that God created the Earth as it is in the course of seven 24-hour days. There is no place in this thinking for evolution, no place for a progression of life forms, no place even, it seems, for surprise.

In all of this discussion about the evolving universe, we're talking finally about the truth. We are searching for the truth. We're not talking about what I believe to be the truth. Just because I believe it doesn't make it true. We're talking about actual reality—truth.

The truth is bigger than any human mind can comprehend. It is surely bigger than my mind, bigger than the mind of current scientific thinking, bigger than the mind of biblical scholarship, even— and here we would be severely challenged—bigger than those who wrote the Bible, yes, even than the Bible itself. The Bible is not God. Only God is God. And God is the Truth.

Many times in my life I have had the distinct sense of being led by unseen forces into ways that I did not understand and with which I did not even feel comfortable. Looking back on those times I have generally realized that I was being led to a truth that was much bigger than I. Those times are unsettling. But they are of God. They lead toward the truth, which is always there at the end by surprise. It never has needed me to defend it or even to explain it. The truth could always take care of itself. The truth was of God. The Truth was God.

So must be our search for the truth about the universe and our life in that universe. There are many trustworthy guides—science, the Bible, our reason. We must be open to the revelation of the truth, come from where it may and by whatever means. For it is the truth.

Edward O. Wilson, the Harvard biologist, in his book, *Consilience*,[11] makes an articulate case that science and the humanities must come together in the quest for the truth. He sounds a strong alarm that if all disciplines don't work together, the world is in for catastrophic changes due to humanity's present destruction of the environment.

There is also a great need in our day for theology to tie together creatively the emerging scientific record of the unfolding of life in the universe with the biblical record of the relationship of God with humanity. Can it be clearly seen that the God who led the people of Israel out of bondage in Egypt is the same God that saved the early bluegreen photosynthesizing bacteria through the emergence of the breather bacteria? Who can show us that the God who raised Jesus from the dead is the same God that raises new worlds out of dying and collapsing old stars?

Barbara Brown Taylor, in her small but stunning book, *The Luminous Web*,[12] makes an inspiring case for theology to work with science in the pursuit of the truth of the universe. Truth, in the end, she states, is mystery for both disciplines.

The four Dynamics—Being, Belonging, Becoming, and Surprise—are offered as a framework to make that connection between theology and science. Theology must weave them into the biblical fabric and the scientific evidence.

THE DAWN OF HUMANITY

But we still haven't finished with the story of the universe. Let us pick it up 65 m.y.a. It must have been quite bleak after the dinosaurs and most of the rest of biological life died out at that last great extinc-

---

[11]New York: Alfred A. Knopf, 1998.
[12]Cambridge, Boston, MA: Cowley Publications, 2000.

tion. Lonely winds whistled down empty canyons. Waters fell with no ears to hear them. The sun beat down on barren terrain.

Except it wasn't completely barren. At night, tiny little rodentlike mammals scurried around. Insects of various sorts continued to buzz. Bacteria ran their cycles, as they had for billions of years. Plants and trees soaked up air, water, and minerals. A few lizards basked and some birds flew. Fish plied the seas and lakes.

Into empty ecological niches, with an explosion of creativity, spread the mammals. From being a small night-feeding band under the shadow of the mighty dinosaurs, the mammals quickly branched out into many forms and adapted to the feeding and nesting spaces vacated by the dinosaurs. Some of those early forms would hardly be recognizable now—giant six-horned saber-toothed plant eaters, piglike mammals with saber teeth, smaller animals with the body of a deer, the head of a giraffe, and horns, horses with claws. In many ways early mammal life resembled a Dr. Seuss book more than present reality.

But there were also very familiar forms—turtles, crocodiles, and, later around 30 m.y.a., rodents—arguably the most successful of all mammals. At that same time, there was a marked change toward a cooler and drier climate. This resulted in a retreat of forests and the emergence of grassland savanna and prairies. During this time, all modern hoofed animals evolved. In fact, a greater variety of mammals existed then than at any time before or since. Elephants, apes, and rodents all flourished.

But the climate continued to deteriorate, culminating in the ice ages of the last two million years, lasting up until just 20,000 or so years ago. Remarkably, during the ice ages some of the largest species of mammals flourished. Giant forms of elephants, mammoths, horses, bison, wolves, deer, and cheetahs arose only to go extinct at the end of the ice ages and leave their smaller relatives to inhabit our world.

Meanwhile, a similar revolution swept the plant world. Flowering plants had started 110 m.y.a. but had been relatively rare until after the Cretaceous extinction of 65 m.y.a. Then they spread rapidly, until they now make up 90 percent of all land plants. The explosion of flowering plants occurred because pollination through animals and insects worked much better than wind-driven pollination. So life favored it.

A corresponding increase in insects made the ecosystems work. More particularly, social insects—ants, bees, wasps, and termites—developed. These social insects almost took over the world and, particularly my house and yard. A battle to the death is still being waged between them and us humans.

This last age is the Cenozoic Age, the age of mammals, flowering plants, and social insects. It is the stage for the emergence of the first humans just a short time ago. If my trip to the San Diego Zoo is any indication, those humans might mark the end of the Cenozoic Age. But we rush our story. Let us now look at those humans, their past, their present crisis, and their future possibilities.

CHAPTER 4

Humanity

THE HUMAN RACE is an extraordinary development for life on this planet. What we have come to, in our no more than five million years here, makes a remarkable story. But the way we should relate to the Earth and the rest of life is still not settled. Take that day on Hatteras.

It was the hottest day of July when we visited Cape Hatteras, the easternmost part of North Carolina. Despite the constant sea breeze, the air blanketed my wife and me like a sauna, and the sun battered us with an intensity that could barely be relieved in the shade. Carol chose to lounge around a small pond separated from the ocean by some sturdy dunes. I, always wanting to assert myself beyond what reason or nature would suggest, headed out a couple of hundred yards to the point of Hatteras.

At the point, two currents contend for space. The Gulf Stream comes up from the south with its warmth and builds the southern flank of Hatteras. From the north comes a branch of the Labrador Current, or perhaps it is just a cooler back eddy of the passing Gulf Stream. Whatever it is, it comes from the northeast and builds the east side of the cape. Under certain conditions, one can actually feel the difference in the temperature of the water between the eastern and southern sides of the point.

Out at the point, the two waters clash, sending splashing waves into the air as if they were crashing on rocks when the only thing there is the "rock" of the other current. It is worth visiting. So I trudged along the beach despite the heat in order to witness the clash of the titans.

Standing at the point I acknowledged once again that nature with all its power is there, there, there. We, mere creatures of nature, can only stand in awe and wonder, marking it with our admiration.

After a while it was time to return, for Carol had probably been waiting long enough. So, despite the heat, I headed straight back toward her pond, cutting across the hot sand rather than walking back along the cooler water. The most direct route lay through a little grassy area which, as I got closer, I could see was ringed with signs spaced every 20 feet or so.

The printing on the signs became legible about the same time the birds attacked. The signs said, "Tern Nesting Area—Keep Out." The attacking terns were oblivious to the fact that I could read the signs and that this was a sign I would obey. Some 30 or 40 of them swooped down through the heat directly at my head with a raucous chatter punctuated with loud shrieks as they screamed by. Again and again, they swooped as I gamely trudged on.

Not knowing the limits of these birds, I put my camera case on top of my head, just in case they might misjudge a swoop. Still they screamed at me for coming near their nests. I tried to look insignificant, uninterested, bored—anything to show them that I had no designs on them, their nests, or any eggs that might be there. Finally, I cleared their area, and they lost interest in me.

Several things were quite clear. First, they had had definite designs on my head, for my camera case was scarred with bird droppings. Better the case than my head, I guess.

Second, it was clear that they would surely prevail if it were just they and I. If I would stand a chance against their fury, I would have to bring in reinforcements such as bulldozers and asphalt. That day on the Cape, it struck me with the force of a blast of heat that there were places for them and there were places for me. What was finally at stake was not just their place or mine, but rather *a way to live together.*

Looking at the larger picture, if I wiped them off the Cape and even off the Earth, I would probably never be aware of the differ-

ence. I could do them in, and then the bald eagle, and then the ele-
phant, and so on. But how far could I push my dominance and still
survive myself? How long could I pretend that I was not finally
absolutely dependent on all of these creatures and their vast and
complicated society? How far could I go without a deep acknowledgment that I was part and parcel of the birds, rooted in the life that
is theirs and tied in deep ways to their successful continuation?

WHAT IT IS to be a human being, living with and relating to other
humans as well as the rest of life on this Earth, remains a mystery.
To unravel the incredible greatness and the appalling weakness of
this human creature, we must trace the coming of humankind, *Homo
sapiens.*

We come from the primate line of mammals, starting in the form
of a small tree shrew some 70 m.y.a., five million years before the
extinction of the dinosaurs. This small, rather insignificant cluster of
species lived out of harm's way in the lofty crowns of jungle trees.
About 40 m.y.a., the anthropoid line of primates, leading directly to
humans, branched off. These creatures all lived in the trees, feeding
on nuts, fruit, and berries.

Then, about 20 m.y.a., during a general cooling of the worldwide
climate, Africa collided with Eurasia in a mighty grinding of continental plates. The brunt of the collision occurred in East Africa and
resulted in a marked elevation of the land. This produced a definite
cooling there so that open woodlands quickly replaced the dense
forests with their abundant food supplies. One ecological niche was
closing, and a new ecological niche was opening. A new, more
humanlike branch, the hominoids, filled the new niche. They came
down out of the trees on occasion, learned to eat tougher nuts, and
quickly spread over a wide area of what is now Africa, Europe, and
Asia. About 13 m.y.a., as part of this development, the branch leading to orangutans split off from that leading to humans; at 8 m.y.a.
the branches leading to gorillas and humans split.

About seven million years ago, more or less, another climatic shift
occurred, and the Earth got cooler and drier still. The effect was a

further recession of the jungles and woodlands and the appearance of vast grasslands and savannas. This opened another niche, and to fill it our direct ancestor, the hominids, split off from the branch that would lead to chimpanzees. Coming from a common ancestor, chimpanzees are still our nearest relatives among primates, having DNA structures that are remarkably similar to ours.

These early hominids were definitely bipedal creatures, walking on their hind legs so they could see over the tall grasses and so their hands would be free to gather the new kinds of food to which they were adjusting. Once in a while, these early gatherers would come across an animal carcass left by another predator and eat it with their bare hands. A prevalent theory holds that about 2.5 m.y.a., these early humans developed the first tools in order to cut up the found meat rather than just tearing it. It wasn't until much later that we see hominids actually engaging in hunting in a systematic fashion.

The fossil record reports a growing brain in these hominids around 1.5 to 2 m.y.a. This significant change laid the ground for all that was to come. But what was the pressure for the development of this big brain? Scientists have a number of theories, but let me point out several that make sense to me. As climates and food sources changed, there was pressure to eat meat—not just on occasion, but as a regular part of the diet. With nothing but hands and feet, those primitive people had to develop new tools for the actual hunting. Hands alone were not enough. Brains for tool production would be a big help. Again, food supply drove evolution, although that adjustment could be attributed to the competition between individuals for good food supplies.

Still, there appears to be one further reason for brain growth. To be a gathering society, the women had to carry their young babies with them as the tribe roamed from area to area. This produced a strong mother-child relationship, stronger and of longer duration than in most species. This bond probably formed the early structures of a society. Then, as meat became a larger part of the diet, there was pressure to share the occasional kill among other members of the clan. Working out these arrangements took brains then, even as it does today, giving further impetus for a larger brain. Archeological evidence points to these food-sharing societies at least 1.5 m.y.a.

Neanderthal, a hominid with a brain even bigger than ours, lived from 130,000 years ago to about 30,000 years ago. In Neanderthal we see the first evidence of spiritual concerns. We will discuss that later in this chapter.

Forty thousand years ago, a fully modern human called Cro-Magnon man, after the fossil discovery in the Cro-Magnon caves in France, roamed Europe. Cro-Magnon is the first evidence of *Homo sapiens*.

Language seems to have been fully developed at this time also. The cave paintings of southern France were done about 30,000 to 10,000 years ago and surely reflect a sophisticated means of communication. We also know that the land bridge across the Bering Strait was open at about this time, allowing these early people to come into North America, where they became the first human settlers, Native Americans or American Indians. But the physical and spiritual developments of Neanderthal, over 100,000 years ago, point to language and thought patterns almost as developed as those of Cro-Magnon. Whatever the timing, the progression of tool making and language abilities led toward big-brained *Homo sapiens*, man the wise, modern humanity.

THE GENETIC PROCESS

Having examined the process by which humanity evolved down the mammal line, let us now look more closely at the mechanics of that descent. Could the actual evolutionary mechanics give us a clue about ourselves and our situation? The answer seems to be affirmative. Stephen Jay Gould in his book, *Ontogeny and Phylogeny*,[13] makes an impressive case that humanity carries the genetic structure of the juvenile chimpanzee within itself. That has some incredible implications.

The difference between the genetic structures of humans and chimpanzees is very slight. They are 99 percent the same, showing more correspondence than do other sibling species who are physi-

---

[13]Cambridge, MA: The Belknap Press of Harvard University Press, 1977.

cally barely distinguishable from each other. What, then, accounts for the considerable difference between chimps and humans?

The difference is in rates of development. Human development is retarded. Now that doesn't mean humans are retarded, even though we sometimes might want to think that. It means that a human develops much slower. It takes a human a full 20 to 25 years to do what a chimp does in just a few years. This means that each human feature developed during that period has a longer time in which to mature and can thus develop much further than a chimp's, and so be very effective.

Take the brain, for instance. The human has more time to grow its brain before the programmed genetic signal to stop brain growth comes in. Most mammals have fully grown brains when they are born. Humans don't finish brain growth until early in their third year. Humans have the longest relative infancy of any form of life. Humans spend fully 30 percent of their lives growing to full stature. Human life progresses from birth to death much more slowly than a chimp's.

It is interesting to look at the physical development of the chimpanzee. As an adult, it surely has just a passing resemblance to humans. But when you look at the baby chimp, there is an incredible likeness to a human. This is before the brow recedes and the jaw elongates in the adult chimp. But since the development of the human is retarded, the human skull never moves beyond the shape of the infant chimp skull.

The effect of this is that when humans die, they are chronologically much less developed than most other creatures. Gould goes on to assert that some of the adult skills that have become instinctive in apes through years of genetic development never even activate in humans. Most primates have well-developed instincts to resolve conflict. They will rarely fight each other to the death but rather, through years of genetic development, have learned to signal submission before they are killed. Humans don't seem to have those abilities but will fight, sometimes to the death, often over trivial things.

It does occur to me that the so-called midlife crisis that many men in our society experience might actually be the occasion when that conflict-resolving instinct kicks in. After midlife men often experience a marked lowering of their competitive spirits and an increased need to build relationships, renew ties to families, and avoid unnecessary conflict. Could that be an example of retarded development?

In general, humans take more time to develop the "adult features" of the chimp; they retain the traits of the infantile chimp. Now don't forget, these juvenile features are highly developed. But the bottom line is that humans are genetically built to be childlike, with the wonders and difficulties of that. We might illustrate this form of retardation with the following chart:

Chimp life span

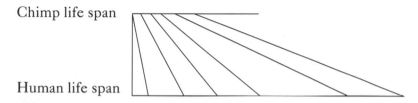

Human life span

This childlikeness in humans is our uniqueness and might be our doom. Consider the features of this childlikeness. Humans are built to love and be loved. Although humans are quite egocentric, they display throughout their lives a ready curiosity, inquisitiveness, and thirst for knowledge. There is imagination and creativity. Humans are spontaneous, playful, and have a sense of humor. Taken as a whole, humans have a refreshing innocence, the freshness of a child. This can be seen particularly in primitive people. As civilization puts on an overlay, it is less clear, but nonetheless true. These genetic traits have much to say about our future and our expectations.

Why then do human affairs often seem so difficult? Why do humans get jealous and mean and fight to the death? Why do humans exploit the Earth, fouling their nest as no other creature does? Is it just a missing gene, or is it compounded by other traits? The answer seems to be that it is both, for there is another very important fact about humanity, at which we will now look.

SELF-CONSCIOUSNESS

The retarded development of humans not only gives the brain a chance to grow so that it can produce tools and make language;

there is an even more radical result. It produces a brain that is big enough not only to think but to think about thinking. We call it self-consciousness.

We have talked earlier about consciousness, noting that it was at least in the earliest life forms in the primal seas and was in potential from the beginning. Consciousness is the ability to be aware of one's surroundings. Consciousness gives an organism, no matter how primitive, the chance to respond to its surroundings.

The big brain of the human has allowed it not only to be aware, but also to be aware that it is aware. It cannot only think, but it can think about thinking as well. It cannot only see into the future, it can see that it is seeing. It is as if we humans stand on a giant stepladder with the consequent ability to look down on ourselves, to look back into history on ourselves, and also to see out forever. We call this ability self-consciousness.

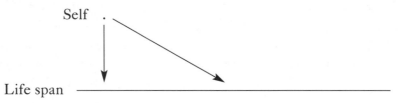

Self-consciousness allows us to transcend ourselves, stand above ourselves and see. That ability is a function of the size of the brain. Humanity achieved self-consciousness when it evolved to having a brain size of about 2000 cubic centimeters.

To our knowledge, the first hominids that had this ability were the Neanderthals. We surmise this because about 100,000 years ago they started burying pots, clothes, and weapons with their dead. They were equipping their dead for a journey to the land beyond. They had the self-consciousness to imagine it.

Neanderthals were the first to speculate there might be something beyond the grave. Not only could they see that they would die, they saw also that there might be something beyond death. They acted as if they were not limited to the eyes in the fronts of their heads, but also had eyes way above their heads so that they could see forever. They could then imagine what might be out there. And they could prepare for it.

This ability of self-transcendence has positive and negative consequences. Positively, with this ability we can anticipate what is coming. This gives us the ability to plan, to create, to build magnificent cities and master-works of art. It allows us to design great societies and produce complex systems. Only a creature with this ability could write a novel. Only one with this ability could absorb itself in the reading of that novel. All altruism, all love, all courage thoroughly interconnects with our self-consciousness. Our brain is finally big enough to do this. It is the single most important characteristic of this new species, *Homo sapiens.*

Now, of course, *Homo sapiens* is known for the opposable thumb, which lets us grasp objects, make tools, and build cities. So when this self-conscious big brain of ours teamed with the opposable thumb, it produced the most incredible species that has ever lived on this Earth. Emerging by evolution from other, less self-conscious species, this remarkable race can communicate in elaborate ways, even around the world in a split second. It has developed incredible technologies that probe into the smallest corners of nature and reach out into the farthest expanses of the universe. This human species has mapped the Earth from outer space, mapped the seabeds from submarines, and mapped the moon from moon orbit and even from the surface of the moon.

The evolutionary leap from the chimpanzee to *Homo sapiens* is as huge as the leap from the first single-celled monera to the relatively gigantic multicelled organisms in the primordial sea. Only when we moved from those early bubblers to the early bluegreens to the first breathers do we see a jump as remarkable as this one.

If that were all, what a fairyland we would live in. But there are negative consequences of our self-consciousness. Thinking about my time on the point at Cape Hatteras, I am driven to acknowledge that. For standing out there, I started thinking about the lovely house I could build on that point. It would be firmly anchored with pilings driven deep into the ground. It would have a sun deck on all sides so that there would be a view in every direction. It would be well air-conditioned because the days can get hot.

There I could have peace and solitude and an unparalleled view of nature at its wildest. On occasion, I could entertain in a fashion that would draw people from some distances. There, I would be able to live in luxury in a manner that would be the envy of many people.

Of course, if I were to do that I couldn't let nature be as wild as it might like. I'd have to build sea breaks and sea walls. I'd have to contour the land. And the terns? Well the terns would have to go. There's no way that they could continue there if I were to live there.

The Cape and the terns are fortunate, however, because the contingencies of the sea and weather as well as the rules of the National Park Service will not let me build that house. But we humans have built many such houses in other places. We build them even when we know that hurricanes and weather can destroy them at will. And we build them even in the most fragile ecosystems. What is it in humans that permits us to do this?

It always goes back to self-consciousness. Our human self-consciousness has so dominated our beings that we tend to be conscious of little else beyond our selves. This big brain is such an overwhelming phenomenon that we put most of our energies into using it and being used by it. It pulls us into itself and focuses our lives on ourselves, to the detriment of all that is around us. We live as if we are unrelated to everything else. We see ourselves as so much more advanced than everything else that we are free to ignore everything else, using it only for our own ends. This self-consciousness even seduces us into thinking we are no longer an animal, but some unique being above animals.

As a result, we have little awareness of our relations to anything outside of our selves. It is as if each of us were the center of all that is. Denying our relatedness to everything else, we replace it with what seems to be an unrelenting greed that threatens to consume everything on the Earth—land, minerals, wildlife, air, and even, maybe, in a grim irony, ourselves. It is clear that the creation "worked" perfectly well before we and our brain came along. It is not yet clear what the effect of this new phenomenon will be. From all evidence it will not be as it is now.

A friend of mine talked about a personal struggle to get beyond his brain domination and listen to his intuitive self rather than his mind. Now this man is naturally quite intuitive as well as having a healthy understanding of himself. But he reported what many of us have experienced. No matter how hard we try to listen to our hearts, our minds continue to chatter, thoughts rambling through them in unconnected ways, ideas and worries popping in and out in a seemingly random fashion, plans, hopes, ambitions, fears always stirring, never quiet.

But would it make a difference if we were able to turn off our big brains? Would intuition by itself make us less rapacious? Many people in our day romantically pursue that course through a number of anti-intellectual movements. In balance, however, it is a rather academic question, because we do have the big brain. It is a present reality and is thoroughly intermixed with our intuitions and feelings. There is no getting around that reality.

Is it just the size of our brain that is the problem? Ernest Becker in his book, *The Denial of Death*,[14] helped me understand that the *fear* of death, occasioned by our big brain, is the actual culprit. He persuasively argues that my self-consciousness allows me to see that there was a time in the past before which I wasn't, and there will be a time in the future when I will be no more. Even when I am not directly threatened, I know I will someday die. By being able to transcend myself, I can see that I am finite, limited. That realization prompts a visceral, intuitive, feeling reaction that generates the problem. Our brain and our intuitions are absolutely interwoven into the situation. Together they create the problem.

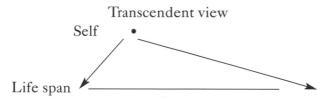

Now these limits don't have to do just with death when the undertaker moves in. They have to do with all of my life. Death is just the symbol of my limits, my finitude. The main point is that I have limits. And I know it. I experience those limits in a number of ways. I study for an exam and then flunk it. I'm just not smart enough. I court an attractive woman and get rejected. I'm just not handsome

---

[14]New York: The Free Press, 1973.

or nice or sexy enough. I apply for a job and don't get it. I'm just not qualified enough. I work in a job for a number of years and am then fired. I'm just not able enough. I want to walk along the beach and the nesting terns will not let me. I'm just not powerful enough . . . unless I bring in the bulldozers.

All through life there are reminders that we are limited, finite. Death is just the final, incontestable fact. It thus becomes a symbol for all of these other, smaller deaths.

The problem for us humans is in the anxiety that comes in the anticipation of death. Otherwise, we would just bounce off life's little brushes with death and need, winning some and losing others. Though not always smooth, it would be consistent with the Earth. The reality is that my anticipation is accompanied by anxiety. Will I win? Or not? Will I die? Or not?

Fear, with its accompanying anxiety, is the almost constant state of our human existence. The juvenile chimpanzee that we are feels very vulnerable and so is anxious. My juvenile chimpanzee wonders if it will have enough of whatever it needs. But my big brain has given me the intelligence to protect myself against danger. I can gorge myself to protect against want. So I build a house fortress on Cape Hatteras. It is more than I will ever need. It is without acknowledgment of any other life around me. It is the product of my self-consciousness.

The problem is not, of course, finally in the anxiety. If that were so, the worst we would have is a slightly quivering Earth, quivering from all of the human anxiety. The problem lies in our reactions to the anxiety. Out of our anxiety, we humans seek to control that which makes us anxious. We dominate, exploit, manipulate, and destroy not only other humans and other species, but also the very life support systems that allow us to live.

If we don't like the temperature the way it is, we change it through heating or air conditioning. If we don't like the way a river runs, we dig it a new channel so that it will run differently. If a particular piece of ground has too many trees, we cut them down and convert the ground to pasture or an apartment complex. If it takes too long to get to a certain place, we'll invent a new and faster way to get there. If something or someone gets in our way, we destroy them. That is ultimate control.

We make control a way of life. This human-initiated control is altering the very basic systems of nature without any guarantees that the new systems will have a place for us. So out of anxiety for our continued existence, we actually threaten our continued existence.

The question can be properly asked: Has it always been this way with *Homo sapiens?* Or is this just a modern aberration? To be sure, some modern cultures, particularly in the East, don't appear to be nearly as anxious and, therefore, as controlling and destructive. That suggests something in modern Western culture makes the problem worse. Nevertheless, it is sure that the writers of the Bible some 2500 years ago saw this destructive nature of humanity quite clearly. The stories of Adam and Eve, Cain and Abel, the Tower of Babel, and Noah's Ark all have a dimension in them about the consequences of human control. What is modern is our particular culture and our technological prowess, giving us the means to destroy the creation on a wholesale basis. It is rooted in our self-consciousness.

Now let us compare us humans to the rest of the animal world. Every conscious creature on the Earth experiences limits. A snake on the hunt for food quickly knows whether or not he is big enough to swallow a certain prey he sights. Most of this is instinctive and is a bodily reflexive response mediated through the nervous system. But some of it comes by thinking in a dull-witted way. Animals are driven mostly by whether or not they are hungry or whether or not they are hunted. They kill animals of other species when they need to eat, and they run when they are to be eaten. It is instinctive.

Within their own species, other animals rarely kill one another. They rarely take more than they need. There is, of course, conflict. It is over food or mates or nesting areas. And the conflict is sometimes to the death. But generally there is an instinctive way to resolve that conflict, to declare winners and losers. An animal will go belly up or will put its tail between its legs or use some other signal to end the conflict short of death. So why can't humans do that?

When it comes to basic instincts, there seem to be two major differences between humans and other animals. First, according to Stephen Jay Gould's hypothesis, the adult instinctive responses bred into mammals over millions of years of genetic development have not been passed on from the chimp to the human because of the process of retardation that we spoke of earlier. We humans don't instinctively know how to capitulate before danger. It's not in our genes. It is not a part of the process of our retarded development.

The second major difference is that our big brain and its thought processes has taken over many of our responses to life that were previously handled by instinct. The human brain is so active and has so much imaginative power that bodily instinctive reactions have little room. Instinct has been overwhelmed. To raise a child in our culture, we must have the right baby books, which instruct us how to react to the baby according to the latest theories. Even sexuality must be taught. Obviously, we can't do without training for child rearing and training for sexuality. But in the past both have been deeply instinctual functions. Now we need to be taught because our instincts have been replaced by our minds.

This big brain of ours has not only limited our use of basic instinct as a guide to navigating life; it has been the occasion for us to imagine way beyond our immediate experience. We can see possibilities. We can foresee, even imagine, danger. Through our brain, we construct grand developments because we can imagine them. Because of our brain, we react to danger with anxiety. We seek to avoid the danger through elaborate schemes of control whose by-products now threaten our very existence.

In a letter to me, Dr. Danny Martin eloquently paraphrases Ken Wilbur[15] and gives a slightly different slant on this. He says,

> *Sometime during the second millennium BCE (Before the Christian Era), from the world's myths a cry of guilt and sorrow screamed out as humankind finally emerged into the light of consciousness and faced for the first time the frightening spaces of a new, mortal and lonely existence, no longer protected by the womb of the subconscious. However, this tiny breakthrough immediately began a project of its own: by attempting to deny its roots in the subconscious and by despairing perhaps of the future, our ancestors began to remake the cosmos in their own image; a protest against the Earth that in time became an assault on nature. What Tom Berry calls the great rage of the human species against its condition.*

---

[15] *Up From Eden* (Boulder: Shambala, 1983).

Whatever the reasons for our actions, we are in a dilemma.

The consequences of this "condition" can be seen as nations seek to defend themselves against enemies. During World War II, the United States was quite sure that Germany had the skill and the brains to build an atomic bomb. So the United States set out to beat them with a massive crash program, the Manhattan Project.

The war with Germany ended, and there was little danger that Japan would develop the atomic bomb. Nevertheless, the United States proceeded to drop two bombs on live Japanese cities. Even before the first test of the bomb, there was talk that it might be powerful enough to ignite the atmosphere and bring the world to an end. But in what seems a massive denial of relatedness or consequence, the developers persisted and ran that first test.

Of course, there were valid scientific and strategic reasons for both doing the test and dropping the first bombs. The internal debate was fierce, with good arguments on both sides. Nevertheless, the consequences are that humanity now has the power to destroy itself and much of the rest of the creation.

Self-consciousness, despite a great power advantage, could only lead us to imagine ourselves without power and in need of ultimate power like the bomb. Self-consciousness drew all our attention to ourselves to the exclusion even of nature. National insecurity and personal dissociation changed the world forever. We don't know the ultimate results of those decisions, even to this day.

Will it ever be possible for humans to react to the world—either the desire to build on Hatteras or the necessity to defend against enemies—from other than a narrow personal point of view, a point of view unrelated to the rest of the creation? Will it ever be possible for humans to see themselves for what they really are: animals with limits and needs and related to everything else? Will we ever be able to see ourselves essentially as part of a larger, interrelated system?

SIN AND EVIL

None of this is new to theology. In theological terms, we call our self-centered, unrelated response to threat sin. Sin is making ourselves number one, before everything else. It is refusing to value first the community of which we are a part. It is putting ourselves in the place of God, acting as if we were God, setting ourselves up as a little tin god where everything revolves around us. So when anything threatens us, we lash out at it, seek to control it.

We sin when we try to protect ourselves against a perceived danger without taking anything else into consideration. I cheat so I won't fail the course. I lie so that I won't be punished for what I have done. I steal so that I won't run out of money or so that others will think I am important. I will even steal the nesting area of some terns so that I can have a beautiful homesite to enhance my sense of self-importance. I will kill not just so that I won't have to be killed myself, but at times, so that I won't even have to be inconvenienced. Any sin we can think of comes at the occasion of knowing that I am threatened and might lose, knowing that I am threatened and wanting to appear not threatened.

There is another possible way of looking at sin: the refusal to be who we were created to be. When we look at humanity as descended from the juvenile chimpanzee, we can see sin as a response to life that is not congruent with our ancestry. Rather than being playful, loving, creative, or inquisitive, we sin when we seek to control, dominate, or protect at the cost of all others. When we view ourselves in evolutionary terms, we could even speculate that sin is our attempts to be an "adult" when we don't have the genetic, instinctive equipment to do that. Sin is the refusal to be what we were created to be— loving, open, vulnerable, related.

Whatever the source of the sin that enshrouds us, humanity seeks to build a civilization where there would be protection from the whims of nature, disease, and other humans. What we get instead is a civilization that could be the last one for humanity. We erect a complicated technological system that has a solution for every ill and yet leaves in its wake a land that is becoming unlivable because of pollution, toxins, and mounds of garbage. We arm ourselves, searching for protection, shooting and being shot in the name of that protection, yet leaving everyone's lives even more precarious from a plethora of ways to kill each other. We war and threaten war, always in a bid to make the world safe from war and seeming to leave it always more vulnerable to war.

Sometimes it all seems so overwhelming, like we are caught in a web of sin and evil, where we are not just the perpetrators but also the victims. And the harder we try to get out, the deeper we get in. The situation is further complicated when individual acts of sin become institutionalized into larger systems of unrelenting evil— Nazi fascism, racial bigotry, sexual discrimination, corporate greed, contempt for the poor, spiraling arms races and wars. Aren't these the principalities and powers spoken of in the New Testament?

For the purposes of our present discussion, we must ask if these evil systems are just part of the human condition. Or are they innate to the very creation itself? Is there an objective force of evil outside of humanity that has entrapped the creation and from which humanity, and the whole creation, needs redemption? Is it just humanity that is fallen, or is it the entire creation? How can any understanding of sin and evil be reconciled with a growing understanding of the evolution of the Earth?

When we look at things from the evolutionary perspective, the creation is not fallen—it is natural. There is an ecological balance that supports life. To be sure, there is danger and destruction within the creation. Nature is "red in tooth and claw." From the point of view of any one species as it is attacked by its natural predator, the world might seem possessed by "the evil one."

Seen from the broadest evolutionary point of view, however, the only "fallen" thing is humanity. Humanity is what is tilting the Earth system so out of balance. That fallen state of humanity is the result of a big brain that can see forever and so worry about its end. That fallen state is the result of a youthful genetic track that wants to play and create but has no instinctive responses that know how to deal creatively with need, danger, and conflict. That fallen state drives us to control and kill, so that we will not die.

On the other hand, we must admit that if humanity can be evil and humanity is part of nature, then there is evil in nature. Can we ever see evil beyond human activity? Perhaps. Cancer, for instance, seems utterly devoid of any sensibility about anything beyond itself.

It is self-serving and self-referencing. Could cancer be a reflection of a larger evil force in creation?

Another possibility suggests itself. Could evil be present in every species at its time of speciation, before it reaches its new state of equilibrium? At the time of speciation, there is swift change and a certain "anxiety" present. Perhaps that "anxiety" is the occasion for "evil," as far as that species is concerned.

These possibilities could provide the natural basis for the long and honored tradition in theology that sees evil as an objective force within the creation. That tradition sees evil as demonic, having humanity in its grip and bent on its destruction. Evil is symbolized as the work of Satan or the Devil. Is there a way to correlate this traditional understanding of evil with an emerging creation? Where could evil be rooted in the creation? How could it fit with the Earth before the advent of humanity? How would nonhuman forms fall within its sway? How could we understand redemption from evil for the amoeba, the deer, and the caterpillar as well as for humanity?

Finally, no matter what theory of evil we accept, it is sure that humanity thoroughly participates in sin and evil. It is at the root of our self-destructive course. We must be delivered from sin and evil if we would survive. How will we escape? To paraphrase St. Paul, "Who will rescue me from this body of death?" (Rom. 7:24).

THE SPIRIT

Now we must touch one more base to get a clear picture of this emerging creature, *Homo sapiens*. We must talk about the spirit. We know that reality is not just tangible, that which we can relate to through our five senses. There seems to be another invisible, inaudible, untouchable spirit world to which we can relate by means of a "sixth sense."

Science and religion have each sought to understand this spirit world. Scientists do experiments on extrasensory perception (ESP) and parapsychology. There are substantiated examples of levitation and other phenomena that defy traditional scientific understanding. People with special psychic powers are sometimes even brought into police cases to help locate missing persons or provide leads to puzzling crimes. Some science seems to acknowledge the existence of a world that operates beyond our regular five senses.

The Institute of Noetic Sciences (IONS)[16] is an organization founded by Astronaut Edgar Mitchell. Mitchell had a remarkable experience on his trip back from the moon in *Apollo XIV*. It was an experience of expanded consciousness, where he deeply experienced belonging to the whole universe. Trained in the hard physical sciences, he set out to discover the connections between that physical world and the inner world of subjectivity. A secular organization, IONS currently sponsors research, conferences, and training to explore the connection between the world of the spirit and the physical world.

On the other hand, religion has long seen the spirit world as its special domain. Prayer and spiritual healing are commonplace. The spirit world is the place where much religious thought insists that God dwells. For traditional religion, it is through spiritual means that we can best make contact with God.

The afterlife also draws the attention of religion. Speculation about conditions in the next world ranges from choirs of singing angels to an everlasting golf game. Some see Heaven, Hell, and Purgatory as the basic areas of that spirit world. Others are less descriptive, or speak of climbing through a series of learnings toward heavenly bliss. Just about every religion has some idea about what one needs to do on this Earth in order to get to Heaven.

Those who take an evolutionary point of view are apt to say that evolution has now advanced far enough for humanity to have one foot in the spirit world. They would say that we are evolving in the spiritual direction. It is as if we were built to leave the physical and continue in the spiritual world.

To be true to everything we have said previously, we must insist on one thing: the world of the spirit is no more or less the domain of God than the material world or the world of biological life. God is the God of all of reality—material, spiritual, biological—everything.

---

[16]Institute of Noetic Sciences, 101 San Antonio Road, Petaluma, CA 94952–9524. www.noetic.org. Phone: 707–775-3500.

The modern heresy, which has gotten us into such environmental trouble, is that there is a dualism in the creation in which the spirit world is superior to the material world. If the universe is one interconnected reality with one God, then the spirit world and the material world are one. They are interconnected, intertwined, coming out of one God and, therefore, of equal weight, value, and significance. To construct any hierarchy of value is to dishonor God, the Creator.

We will now leave speculation of the place of the spirit world until we talk later in Chapter 6 of a new way to understand the place of humanity in the creation. First, we must consider the place of Jesus, the Christ, in this long drama. For, as Christians, we are convinced that his contribution is central.

JESUS, THE CHRIST

Jesus, a Jew living some 2000 years ago in ancient Judea, stood in a long line of the people of Israel who had a special understanding of the ways of God and humanity. But Christians see Jesus as much more than that. They see him as the incarnation of God, come to establish an even more perfect way for humans to relate to God. That new way was inaugurated in his birth and baptism, conveyed through his life and ministry, and sealed in his death and resurrection.

Jesus came to show us what it means to be human, fully human, as we were created to be. A professor in seminary once said to our class, "Don't ever let me hear you say, 'I'm merely human.' To be human is what we are created to be. To be human is our goal, our calling. But our sin pushes us to be either less than human or pretend that we are superhuman." The goal in life is to be human to the fullest. Our present crisis comes in part from not understanding that. Jesus is the model for our humanity.

So what does it mean to be fully human? If the four Dynamics describe the way the universe works, then humanity, at its fullest, must live the four Dynamics. If Jesus is the model for what it is to be human, then Jesus embodies the four Dynamics. In other words, the universe best expresses itself through humanity in Jesus.

Let us then look at Jesus and the four Dynamics. First, the Being

Dynamic. Jesus was whole in himself—perfect being. He had an integrity within himself, an authenticity that drew attention. Even his detractors commented on numerous occasions that he spoke with authority. Jesus was the embodiment of the Being Dynamic. He set the standard for what it is to be a human being.

The Belonging Dynamic is closely related. To be fully human, not only must we have integrity within ourselves, we must at the same time belong to the greater whole. We are part of the human family, part of the Earth, part of the universe. Jesus was the Belonging Dynamic in human flesh. He spoke of loving not only our friends but also our enemies (Matt. 5:43–48). He lived out a love that insisted on belonging even to the extent of losing his life, his own being, to his belonging. Jesus lived a love that was sacrificial.

Jesus was quite clear that humans are to be who they are created to be and, at the same time, are to belong to the larger reality of the whole Earth. Being and belonging are interwoven and both necessary for life.

But Jesus was the other two Dynamics also. Life is constant becoming as the ongoing mystery emerges, often by surprise. First, Jesus was the Becoming Dynamic. This is seen chiefly in the incarnation. Jesus was God coming into the world. Take that wonderful passage from the prologue of the Gospel of John:

> *In the beginning was the Word, and the Word was with God, and the Word was God. He was in the beginning with God. All things came into being through him, and without him not one thing came into being. What came into being in him was life, and the life was the light of all people. The light shines in the darkness, and the darkness did not overcome it.... The true light, which enlightens everyone, was coming into the world. He was in the world, and the world came into being through him; yet the world did not know him. He came to what was his own, and his own people did not accept him. But to all who received him, who believed in his name, he gave power to become children of God, who were born, not of blood or the will of the flesh or of the will of man, but of God. (John 1:1–5, 9–13)*

The becoming nature of Christ shows itself in the incarnation of God as the human, Jesus. That becoming continues as we look for the second coming of the Christ at the end of the age.

Finally, the Surprise Dynamic is the punctuation of the Becoming Dynamic. Surprise is becoming happening in unexpected ways. The resurrection of Christ is the central exhibit of surprise. Who would have thought of it, much less thought it possible?

Surprise, as an experience of becoming, expresses the intertwined natures of life and death. Life comes out of death, just as death follows life. Again and again in the life of Jesus, just as in the life of the universe, life lived faithfully ends in death, and death entered into by faith is transformed into life. At the heart of reality, the beat of the universe, it is Christ at the core of all that is.

Jesus did not just live as a Jew some 2000 years ago. Jesus, the Christ, was from the beginning of the creation. According to the faith as expressed in those first verses of the Gospel of John, Christ was the very instrument of creation. Thus the four Dynamics found throughout the universe are the same four Dynamics found in Jesus. He is the embodiment of the Dynamics. Their presence throughout the universe story is the trail of Christ, the Creator. Christ's personality, his fingerprints, are evident everywhere in the story of the universe.

What is finally possible here is the reunion of science and religion. It does not have to be two stories, but can be one story. The story will not only describe the external events of the emerging universe; it will describe the internal drama as well. The events will have soul. The spirit will have form. At the center of the story will be the Christ. But that Christ can also be the end of the story, the key to our living on the Earth, as we will discuss in the final chapter.

A SUMMARY

Now let us summarize one final time the way all of this comes together. We humans are the product of the creation, tracing our lineage back to the Big Bang beginning. Most recently, we emerged from the infant chimpanzee and are coded to play, to create, to love, and to wonder. But we have this oversized brain, which is our marvel and our peril. With it we can dream great dreams, and with it we can become terrified and turn our terror into destruction. With it we should be able to live in harmony on the Earth.

Jesus, by embodying the four Dynamics in the form of *Homo sapiens*, stands as the capstone of this long journey. He is the key not only to each individual's life but also to sustainable life for all of humanity. He is what it is to be human on this Earth. He is the fulfillment of evolution, the prototype of the new humanity, which can inherit the Earth if it but decides to. Jesus is, finally, God's demonstration that humanity is fundamentally part of the Earth, destined to voice the praise of the Creator on behalf of the whole creation.

CHAPTER 5

So Here We Are

HUMANITY IS POSSIBLY 40,000 years old as a species, and all is not well. Our relationship to the Earth is troubled. Perhaps nothing describes it so dramatically as Love Canal.

The bus was taking us from a conference we were attending in Buffalo, New York, to the Love Canal neighborhood in Niagara Falls, some 20 miles to the north. It is a nice area, northern New York State—rolling hills, lots of open space, small farms, and smaller towns. We chatted amiably on the bus and did not think much about where we were going.

Presently, we picked up several people who had once lived in Love Canal and now worked with a local advocacy group to keep the cause of Love Canal alive. The woman who spoke for the group gave us a little background. It seems that a large chemical manufacturer, Hooker Chemical Corporation, had dumped about 22,000 tons of toxic chemicals into an abandoned canal, ironically named the Love Canal after its nineteenth-century builder, William Love. In the 1950s the pit was covered with dirt. Then some homes were built on it, and even a school. But by the early 1970s, the wastes were found to be seeping through the ground and draining into the nearby residential community. They were found in the school playground, in home backyards, and even in the basements of homes.

High rates of miscarriages were noticed as well as high rates of cancer. So the homes were ordered evacuated. The dump was covered with clay and roughly 200 homes were demolished. That land is not available for resettlement.

About the time the briefing was finished, we pulled into the community. It contained the modest houses of blue-collar workers. There was nothing fancy. It was pure working-class America. The tranquility was interrupted only by the boards on the windows and doors of many of the houses. House after house was vacant with an official warning sign posted. It was as if a plague had swept through and left an eerie silence in its wake.

Then we spotted the "canal." It was a six-block-long mound over the old canal, now fenced in, planted in grass, again with large warning signs at regular intervals. Everywhere it was grass, green and manicured. At intervals were standpipes through which readings of toxicity could be made.

What was striking was the benign feeling about the whole scene. It seemed so peaceful and pastoral. As we drove around, however, we heard the stories of the deadly poison that was concealed by the green. Here a family had all died. That house had been torn down because it would never be habitable again. We stopped once, and our guide pointed to the six or eight houses in our sight. Every house had been declared safe by the government, but every house contained a case of cancer or other major toxic-related disease. The horror started seeping into our souls as the toxins had once seeped into the neighborhood.

Now the government has declared the whole area safe. The remaining houses are going up for sale. But the activists are resisting. Little trust remains. The ride back to Buffalo was more silent, like the shell of the community we had just left.

HUMANITY IS SWARMING all over the planet. As it swarms, it leaves the waste of its activities in its wake. Perhaps the Hooker Chemical Corporation was grossly negligent. Perhaps it was greedy or just ignorant. What is true, however, is that there are thousands of companies like Hooker around the Earth, doing the things that Hooker did. Millions, even billions, of people support the way of life and aspirations for life supported by Hooker and the other companies of the Industrial Revolution.

Take a drive around any city and the surrounding countryside. Look at a huge open area being scraped by bulldozers to clear for a new shopping mall. Look at the raw earth and remember the life that scurried there at one time—microbes and worms in the soil, grass and bushes and trees on top of the soil, birds in the trees, grasshoppers and ladybugs everywhere, moles and rabbits and maybe deer. Life. But now it is barren. It will be replaced by asphalt and steel and concrete. And human beings.

Fly in an airplane and see the pall of smoke and haze over the city. It hangs like a death shroud, holding in heat, cutting off sunlight, concentrating poisons in lungs. Occasionally, a fresh rain or a blast of cold air will clear the smog out for a day or two. The brilliance and clarity of the new day can barely be appreciated for the contrast it poses with the usual dullness.

Drive into the country and see clear-cut logging operations. Know that the cutting is for boards for your house and pulp for your newspaper. Wonder about the life displaced, the erosion encouraged, and the healing of the scars that will be necessary.

Wander in your supermarket and see the antiseptic produce displayed, each apple big and perfect enough for Adam and Eve, each orange unblemished as if fresh from a juice commercial. And wonder about the chemicals necessary to produce the size and the cosmetics.

Look at the bottled water shelf—bottled fresh from a spring high in the mountains as a substitute for our regular water. Deep in our minds we wonder about our water. We fear it has unhealthy chemicals in it, chemicals that come from our way of life. The chemicals we use to grow big fruits and make them look luscious come full circle to make us afraid.

That supermarket is the symbol of consumerism, thought by many to be the economic drive behind the exploitation of the environment. We in the West have a voracious appetite. As Thomas Berry says, we are gorging ourselves on the Earth. The consumer gadgets are fun and helpful. But are they necessary? And at the expense of the creation?

So what should we do? It is sure that we cannot go back to the "good old days" before we embarked on the wholesale exploitation of nature. Those days are illusive, even fictional. Life subject to rampant disease and arbitrary famine is not to be desired. A flight to "nature" will not solve the human enterprise or the environmental crisis.

But at the same time, we must say that our "progress" has not always been progressive. Somewhere during the past few centuries we have gone astray, or we would not be attempting to conquer disease and famine and replace them with extinction. Could it have been just an accident? Or have we been violating the very dynamics of the universe in our development these last years? Have we been violating God? Let us now look at the current situation and then the four Dynamics for an understanding of our situation.

THE CURRENT SITUATION—BIOLOGICAL LIFE

The emergence of humanity is the overwhelming story on the Earth for the past five million years. For a long part of that period, humanity was a small and somewhat inconsequential part of the Earth. But over the past several hundred years the story has changed considerably. According to 1998 estimates from the United Nations, the world population has grown as indicated in the table on the following page.

This is sobering. But the news is not as bad now as it was just a few years ago. Then, the best median estimate for the world's population at the midpoint of the next century was 10 to 11 billion. But fertility rates have been decreasing so that the newest estimate at 2050 is just over 9 billion. Nevertheless, this population is a tremendous burden on the Earth's systems. The growth of our species is staggering.

Equally important as these figures is the distribution of the world's population. In 1950, three cities in the world had populations over 10 million. In 1990, there were eleven. In 2000, 47 percent of the world's population lived in cities. By 2030, that will be 60 percent. In those thirty years, the total world's population increase will be the same number as the growth in urban population. Urban populations in developing countries are growing at 2.3 percent per year while the rate of growth of the rural population in those countries is

Year	Population
10,000 B.C.	10 million
A.D. 0	300 million
1000	310 million
1500	500 million
1750	790 million
1800	980 million
1850	1 billion, 260 million
1900	1 billion, 650 million
1950	2 billion, 520 million
1970	3 billion, 700 million
1980	4 billion, 440 million
1990	5 billion, 270 million
2000	**6 billion, 100 million**
2025	7 billion, 824 million (projected)
2050	9 billion, 300 million (projected)

0.11 percent. Those rural populations still grow almost as fast as the urban populations of first world countries. The runaway growth is thus in third world cities.

Five factors can cut that growth by contributing to a lowering of the fertility rate: (1) adequate nutrition, (2) proper sanitation, (3) basic health care, (4) equal educational opportunity for women and men, (5) equal rights for women alongside men. The first three directly reduce infant mortality. The last two elevate the position of women, including giving them more control over conception. Together, they empower people, giving them more hope that their children will live into adulthood, thus decreasing the pressures to have many children in the hope that some will survive.

At the same time as this explosion of humanity, there comes the decline and even extinction of many of the other species on the face of the Earth. Since my visit to the San Diego Zoo, I have checked numbers on the rates of extinction and found the chart in the zoo to be fundamentally sound. My research has shown that 98 percent of all of the species that ever existed are now extinct. However, there are probably twice as many species existent now as there were at the time of the dinosaurs some 65 million years ago. These present species are much more complex and much more diverse than ever, a hopeful fact for the future of life on the planet. Estimates of the present number of plant and animal species vary between five million and 50 million with 10 million the most frequently cited number. About 5 percent of the present species have been catalogued, making it difficult to make predictions of rates of species losses.

There are two generally accepted methods of arriving at rates of extinction. Some estimates count the losses within a small area where careful research has been done, such as the Hawaiian Islands. Using that limited laboratory, an educated guess states that the present global loss is 1000 species per year with a prediction that it will go to 10,000 species per year shortly.

A second method looks within a limited area at the consequences of habitat loss. It appears that 15 percent of the species present in a defined habitat go extinct if a habitat is cut in half. Relating that rate to the projected encroachment of humanity on the Earth globally projects a loss of 50,000 species per year, or half of the Earth's species within the next 100 years!

We can then line up these various extinction rates as illustrated in the following chart:

Extinction Event or Prediction Source	No. of Species Now	Rate of Extinction	Predicted Future Rate	No. of Yrs. to the Collapse
Dinosaurs	5 million	100/year		45,000 years
Known Losses Method	10 million	1000/year	10,000/year	1000 years
San Diego Zoo	10 million	1000/year	26,000/year	400 years
Habitat Loss Method	10 million	1000/year	50,000/year	200 years

Obviously, the rate of loss of species will not follow a straight line. In general, it will get increasingly fast before it starts to straighten out to a "normal" rate. Within that change, when will humanity become extinct? What will be left? Recovery after the loss of the dinosaurs took only about five million years. Estimates suggest that recovery from an extinction as severe as the one we are entering will take 25 million years.

Perhaps we can change our assumptions radically to present a less bleak picture. We can increase the estimated numbers of existing species to 15 million or cut the annual rate of loss. Any way we cut it, we are still in the situation that is more dire than when the dinosaurs perished over the course of a million years or so some 65 million years ago.

Humanity is fundamentally changing the beings and belongings of the Earth. It is happening through a great crash of life. That crash, right now, is at a rate 10 times the rate during the crash that eliminated the dinosaurs. Other crashes with which we are familiar have shrieks and roars about them. This one is silent, deathly silent. How do we compute it? How does it get our attention? This crash is of historic magnitude. But it doesn't make the newspapers or TV newscasts. What is it like to be in this kind of crash? Just listen. Listen to the silence.

THE CURRENT SITUATION—LAND

Everything throughout the Earth is interrelated in one large system. We can look at pieces of the system and know that they relate to other pieces, sometimes in ways that are not apparent. One surprising connection is the relationship between environmental degradation and poverty. Gro Harlem Brundtland, the former minister of environment and prime minister of Norway, headed the World Commission on Environment and Development for the United Nations. The commission, composed of members from around the world, came to a surprising and unanimous conclusion that poverty and environmental degradation are closely linked. "Poverty itself pollutes the environment."[17] Poverty, and the need for poor nations to

---

[17]*Our Common Future* (Oxford, New York: Oxford University Press, 1987), p. 28.

repay their international debts, forces them to exploit their farming lands and their mineral treasures as well as causes population explosions. They ravage the Earth because they are poor.

Now let's look at the condition of the land. The Global Environmental Outlook 2000 (GEO-2000) of the United Nations Environmental Programme agrees with the Brundtland Commission. Around the world, but particularly in Africa and West Asia, poverty is leading to deforestation, soil degradation, desertification, and declining biodiversity. These poor countries push agriculture into marginal and wilderness areas, threatening to make deserts out of existing ecosystems. Poverty stresses ecosystems and leads to land degradation.

Northern Africa and the so-called fertile crescent of the Tigris–Euphrates Valley of the Middle East were lush heartlands of food production in ancient times. Over the years they have become deserts, demonstrating that the destruction of the environment is not just a modern phenomenon. If those regions did it in ancient times with relatively low populations, then other areas of those regions are even more vulnerable with their large populations.

Forests and tropical forests are in retreat in much of the world. Slash and burn agricultural methods practiced by certain indigenous peoples persist. But regular forest fires seem to be becoming more frequent and extensive, due to changing weather patterns and land use that has made susceptible areas more prone to burning. Deforestation seems most acute in Africa and Latin America. Tropical forest loss is irreversible and brings with it sharp losses in biodiversity as species can no longer hold on with the loss of their habitats.

On the positive side, deforestation seems to have been reversed in North America and Europe although global warming could move the ideal range for many North American forest species some 300 kilometers (185+ miles) north. This would then call back into question whether we had actually reversed deforestation.

Finally, an emerging problem concerns the vastly increased introduction of nitrogen into Earth systems. The World Resources Institute suggests that human activities (overfertilization, fossil fuels combustion, biomass burning, wetland draining, land clearing) introduce 210 terregrams of nitrogen per year whereas natural processes introduce only 140 terregrams per year. Nitrogen is a greenhouse gas so the result is increased global warming. But other serious effects are the degradation of coastal waters and havoc in other ecosystems. Every year there is a dead zone in the waters of the Gulf of Mexico because of the concentration of fertilizers dumped into it by the Mississippi River.

THE CURRENT SITUATION—WATER

According to GEO-2000, water deterioration is a major environmental concern around the Earth. In Asia, water supply is a serious problem. Already one in three Asians has no access to safe drinking water. Rapid urbanization is putting increased stress on water supplies in Africa, Latin America, and West Asia. Poverty, again, is a major factor affecting water supplies, particularly in Africa. By 2025, 25 African countries will be subject to water scarcity or water stress.

Overexploitation of ground water lowers underground water tables, which can damage wetlands, and, in coastal areas, set up the conditions for the seepage of salt water into underground water supplies. In many large cities worldwide, inadequately treated sewage is discharged into ground water reserves, resulting in the lowering of oxygen in these waters, which can cause great damage and even death to fish and other creatures living there. If the resulting polluted water is not adequately treated, then infections dangerous to humans can also result. Runoffs of fertilizer and pesticides in rural areas and runoff of petroleum products waste in urban areas can also damage ground water supplies.

According to a World Resources Institute report, freshwater habitat comprises only 0.8 percent of the Earth's surface, but is home to 12 percent of the Earth's animal species and 2.4 percent of all plant and animal species, making it the richest habitat on the Earth. But freshwater habitats are declining rapidly from such causes as physical alteration (dams and canals) water withdrawals, overharvesting, pollution, and the invasion of non-native species. On the positive side there are moves worldwide to dismantle dams and, in the United States, there are other projects such as a $7.8 billion effort to restore the Everglades in South Florida.

Marine fisheries are being overfished and causing problems, particularly in parts of Asia and the Pacific, North America, Europe, and West Asia. In addition, low-lying regions throughout the world are in danger of rising sea levels due to global warming. Climate change also threatens to alter weather and ocean circulation patterns, which can result in more severe weather and the migration of fisheries and the size and even the existence of marine life.

In addition to the above threats, we continue to generate nuclear wastes with no satisfactory means of disposal. Right now, burial in seabed sediment is being tried. It is claimed that the migration of the seabed is very slow and eventually the seabed will be folded down into the interior of the Earth, which is already radioactive. But can we afford to even experiment, with the quality of our oceans at stake?

Our oceans serve as the lungs of the planet. Impurities are filtered out through the marine life in the oceans. The oceans are the source of vast food supplies. But marine life is much less present than it was just 30 years ago. Some has been poisoned. Some has been overharvested. Some is even extinct. Are we poisoning our oceans, too?

The Current Situation—Air

Global climate change is the biggest environmental problem facing the Earth. The burning of fossil fuels (oil, gas, and coal) energizes the world economy. But there is a price to pay. The resulting air pollution aggravates chronic respiratory illnesses in humans, adversely affects the functions of natural ecosystems, damages buildings through acid rain, and causes disturbances in weather patterns. In cities, ozone is generated from automobile engines and can cause severe difficulty in breathing for more vulnerable people. My own city, Atlanta, Georgia, went through a long period when it could not receive any more federal highway funds until an acceptable plan for the decrease of ozone was filed.

This burning of fuels, particularly fossil fuels, deposits growing amounts of carbon dioxide and other greenhouse gases into the atmosphere. These gases are forming a heat shield over the Earth like a giant greenhouse. It is called the "greenhouse effect." Heat that once could have escaped into outer space is now reflected back at the Earth. The result is an estimated increase in average global temperature of 0.5 degrees Centigrade over the past 100 years. That takes into account the huge buffer provided by the oceans. Oceans absorb a great deal of the temperature rise and dampen the effect of the temperature shifts. When the oceans reach their capacity to absorb heat, the temperature could start to rise even more dramatically.

The Earth's atmospheric system is very complex, making it difficult to project the consequences of the increasing release of carbon dioxide into the atmosphere. Current estimates suggest that the world's mean temperature may rise between 3 and 16 degrees Fahrenheit, depending on the climate model and various positive and negative feedback mechanisms. This could cause a rise in precipitation from 7 to 15 percent, and sea level may climb from four to 40 inches. Obviously, a four-inch rise will not be calamitous except perhaps in stormy weather. A 40-inch rise is a far different thing.

As a result of the greenhouse effect, the tropics could cool some because of increased cloud cover. But that will be offset by an increase of 11 to 16 degrees Fahrenheit in mid latitudes and 16 to 27 degrees Fahrenheit in the Arctic regions. As a result of the rising temperatures the polar ice caps will start to melt. Already there are many reports of the retreat of glaciers and early thaw dates. The resulting runoff will flood many coastal areas and force others to build expensive dikes. It will swamp wetlands and force salt water inland, polluting the fresh water supplies of much of the people of the world.

Global warming also threatens to grossly interrupt current weather patterns. Not only will the average temperatures be warmer, but also the vast interiors of the continents, which are often the breadbaskets of the land, threaten to be much drier. There could be drought where we grow our food. The forests, suited to a particular climate, would no longer be able to live there. They will either migrate or die out. A forest can migrate no more than one half mile per year. But the climate change would be faster than that and there are man-made barriers to migration such as cities. So whole forests could perish.

To be sure, there will be some beneficial effects to global warming. The increase in water vapor and carbon dioxide could stimulate plant growth in certain crops and increase rainfall. The climate in high-altitude regions would be improved and heating costs would be reduced. But this will be offset by higher air-conditioning costs. The predicted rise of 2 degrees Fahrenheit by 2015 will raise air-conditioning loads 10 to 20 percent, requiring new electrical generating capacity with its own environmental consequences.

Add to all of this the ozone depletion from CFCs. Those supposedly benign gases that cool our refrigerators, run our air conditioners, and clean our computer chip boards are actually deadly trouble. When released into the atmosphere, they slowly ascend into the upper reaches, where they interact with ozone molecules. They break down the ozone into regular oxygen and chlorine monoxide. Then the oxygen is stripped from the chlorine monoxide, freeing the chlorine to break down another ozone molecule. This one chlorine atom can destroy up to 1000 ozone atoms before it dissipates.

This would not be bad at all, except that the ozone serves to protect the Earth from harmful ultraviolet radiation. This radiation can not only cause sunburn and skin cancer in humans; it can destroy vegetation and sea life as well. It can undercut the very base of the food chain, a process that should be quite upsetting to us humans at the top of the chain. The irony is that we thought that CFCs were absolutely inert, one chemical that had nothing but positive characteristics!

We must interject a hopeful note here. In 1987 the nations of the world gathered in Montreal, Canada, to draft and sign the Montreal Protocol on Substances that Deplete the Ozone Layer. Most of the nations of the world signed, although the United States didn't. Still the United States and others are moving aggressively to cut the production of CFCs that destroy the ozone layer. This coordinated action has brought great progress, so that it seems at this time that the ozone layer will recover by the middle of the twenty-first century.

Most of these changes are combining to alter the basic chemistry of the Earth. They are doing it quickly. There have been climate changes before during the history of the Earth. Some were caused by meteorites, some by volcanoes, some by continental plate shifts, and at least one by the activity of organic life when the early photosynthesizers produced that deadly level of oxygen we spoke of before. But that took ages, and the Earth's Gaia system had ages to

produce an answer. This shift promises to be quick. It is question-
able how much life can adjust fast enough to survive.

It could be that we are in the middle of another giant extinction
like the one 65 million years ago that erased the dinosaurs and some
90 percent of the other species on the face of the Earth. That is hard
to know. But what we do know is that we are facing a wrenching shift.
It will undoubtedly produce vast dislocations and horrendous suffering. We are turning the whole Earth into a giant Love Canal. All
it needs is signs posted around the perimeter: Danger!

THE CURRENT SITUATION—IN GENERAL

According to a World Resources Institute report, the major cause of
our environmental deterioration is the opposite of poverty: it is consumption. This is especially true, of course, in the developed nations
of the world. The world's developed countries have 16 percent of the
world's population but consume 80 percent of the world's goods
through private markets. This contrasts with low-income countries,
which have 35 percent of the world's population, but consume only
2 percent of the world's goods that are privately consumed. And consumption appears to be going up. Worldwide wood consumption is
up 64 percent since 1961, half of it being burned. Worldwide consumption of meat has tripled since 1961, and the worldwide fish
catch has gone up sixfold since 1950. This has resulted in vast depletion of the world's fisheries, as mentioned before.

The technology for solutions to most of these problems exists, even
though funding is often lacking. But, according to GEO-2000, the primary solutions in the future will involve multilateral environmental
agreements, appropriate regulations and incentives, governmental
policies to enforce better conservation and use of resources, and the
inclusion of environmental clauses in trade agreements. These are all
political responses. Effective political responses need the backing of
the people. But the world's people do not yet really see the problem
and, most important, do not yet see that united and concerted action
on a broad scale is what is necessary. The problem is the same as we
outlined in Chapter 1. It is still basically rooted in the way we, the peo-

ple, think or don't think. It involves our lack of an appropriate cosmology. It involves adherence to the four Dynamics.

THE DYNAMICS—BEING AND BELONGING

The basic Dynamics of the universe are being violated. The biosystems of the planet are composed of beings interrelating to other beings in an incredibly complicated belonging. We, *Homo sapiens*, are one of the beings in that giant system. To suppose that we can change the composition of that system without changing ourselves is folly of the worst sort. Everything is interconnected. All is one.

Every being, once created, is now part of the creation. We have to live with and contend with it forever. Those beings and the rest of creation are not there just to serve us. Other beings, even the ones we have created, such as CFCs, toxic wastes, and pesticides, are now part of our bigger system.

Even that is not the end. The deepest truth is that we humans are not clear about what it is to be a human being. We don't know what our role is, what our function is, what our place is.

Our big brain, with its attendant self-consciousness, has brought us a long way. But it also has us terribly confused. Are we an animal like other animals? Or are we something else with other duties? With our ability to analyze and understand, are we to control and manipulate the creation, or not? Are we to submit to death like animals do, or are we to do everything in our power to forestall it? What are we to be?

Because we are so confused about our own being, we can hardly let other beings be themselves either. Like an unhappy child in a classroom, we persist in making everyone else unhappy also. Harmony is not possible when one part of the system is out of harmony.

To be unclear about our being is to also be unclear about our belonging. Humanity's fantasy that some things will not belong to a larger system is a clear violation of the Dynamics. We cannot create CFCs and expect them not to find something like ozone to which to belong.

Greenhouse gases are going to take their place in the atmosphere and start interrelating with other gases, with radiation, and with heat.

In our day, to think in any way other than larger systems is to not think. Everything belongs, or will belong, to something else.

That is also true, of course, for humanity. We belong to the Earth. It is our home, our system. We are just an aspect of the Earth. Essentially, we are the Earth. Everything we do has consequences for other beings. All of our activities affect the whole system of the Earth. But we humans, in our climb up through evolution, have not yet settled on our true belonging. The unsureness of our being and the unsureness of our belonging go hand in hand.

An important breakthrough in my own thinking came when I realized that humanity is still in the process of speciation. Like any earlier species, such as the trilobites, we are experimenting with various adaptations that will allow us to live in harmony in our ecological niche. As we observed earlier, that process can take up to 50,000 years before the fully developed species then lives relatively unchanged in its niche for up to 15 million years. As with any species going through the speciation process, there is disquiet and perturbation during those early years. For humans it is called sin, a breaking of the harmony, an indication that the offender has not yet properly adapted. Our lack of adaptation is playing havoc with the ecosystem. It might never be the same. We might not survive.

We destroy the environment not because we are inherently evil creatures, though from the point of view of other life we might seem that way. It is not that we don't belong on the Earth. We are the most sophisticated organism ever produced by God through the creation. It is, rather, that we have not yet finished our speciation, our development as a species.

When finally speciated and in harmony with the environment, humanity will have both being and belonging. In our being, we humans will know our role on Earth. We will know what we are to do. We will know what we can take, and what we must give back to the Earth. At this point in our development, we obviously have little sense of that. In fact, most of us are not much concerned for the Earth except for what it can give to us. We seem never concerned for the species, *Homo sapiens,* as a whole. There are times when we are concerned for our family or the nation. Most of the time, we are concerned just for our individual selves.

When we finally speciate and harmonize with the environment, the Belonging Dynamic will be fully integrated into humanity. We will see ourselves as one species among a number of others who are all part of one giant living system. Our response to danger will not be just for self-preservation but will be also for the preservation of the whole. Without the success of the whole, we cannot live either. That is slowly sinking into our understanding. But it is far from being instinctual. That will come when we enter fully into our belonging.

THE DYNAMICS—BECOMING AND SURPRISE

The other two Dynamics, Becoming and Surprise, point more to the ongoing processes of the universe. Full of potential and possibility, this universe is becoming. The creation is pregnant with what is to come. The universe is a dynamic event.

The only stance we can take in relation to this becoming is one of awe and wonder. We can only be slack-jawed at the marvel of it. Just as we would not mistreat a pregnant woman, how can we mistreat a pregnant planet? The becoming of the planet is mystery, and we stand hushed in its presence.

How far from this sense of awe are we modern humans, with our controlling ways! We seek to harness nature, not revere it. We seek to exploit nature, not serve it. It is as if we don't expect God to be emerging through the creation. It is as if we have God safely tucked away in Heaven so that God will not be offended by the way we treat the Earth. We are torn apart by a dangerous one-sided dualism. Our behavior clearly violates the Becoming Dynamic.

The Surprise Dynamic accents the Becoming Dynamic. Surprise comes at the ends of things, when God moves in a new way to promote the becoming of the creation. Those of us participating in the surprise must be willing to trust the process, to let go of control, to go with the emerging order, whatever it might be.

The propensity for us humans to micro-manage the process, to stay in control, to do anything possible to keep things as they are, even to the extent of averting our individual deaths, eliminates the possibility of surprise. That is the object of control—no surprises.

But as we have seen, this propensity for control is destroying the Earth. Almost all of the things we do in a bid for control are destructive to our life-support system on the Earth.

So as Gaia is destroyed, the time for an upheaval destructive to humanity and many other beings gets closer and closer. Therefore, this human propensity for control actually hastens the surprise. God will soon have to move in the creation to bring things back into harmony. So the day of the surprise is coming more quickly than it would have if we humans had trusted more in the Surprise Dynamic. What a devastating irony!

Present human activity on the Earth clearly violates all of the Dynamics of the universe. Until just recently, we didn't know that we were doing this. It has just become evident in the past 20 or 30 years. So much of it came out of ignorance or belief in an insupportable ideal, such as progress. But it is becoming clear that we must change our ways or experience great upheaval, even the end of the species.

Even more troubling, this violation of the Dynamics of the universe is also a violation of God, the Creator of the universe. This violation is opposition to Jesus, who revealed the Dynamics to us in the first place. It is even denial of the living Christ, who by means of the Dynamics empowers the universe on its emerging journey. And we didn't even realize it.

Let us now examine the state of the group that bears the message of Christ in our day, the modern church.

The Church

I was born and nurtured in the church. It is home to me. It has given me incredible gifts of life and renewal. I can remember being confirmed at age 12 and feeling the massive (and very hot) hands of the bishop marking me for life. I can remember the sense of importance as a teenage acolyte leading the procession down what must have been the longest aisle in Christendom. Then in college, I discovered the personal message of the gospel. It changed my life and perspective and I decided to enter the ministry. What a gift!

Through seminary, the church gave me an incredible, awakening education about the deeper things in life and the faith. The faith seemed relevant to everything in life, and the church was the instrument with which we eager recruits for Christ would bring the message. Ordination began that wonderful adventure with the message.

That message took me to places I would never have gone but for bearing this gospel. I was privileged to enter deeply into the lives of my people. I was there for births (even one in a backyard—see Chapter 3), for sickness, for tragic death, and for fulfilling death. I wept over the body of a young parishioner who had committed suicide. I grieved with a family whose six-year-old died of encephalitis. I talked intimately with an elderly person about her approaching death. I have laughed and celebrated with my people at weddings, homecomings, and joyful births. I have counseled people through divorce, sickness, grief, and just plain old neurosis. I have been a faithful conveyer of the love of God for God's people.

The church has led me into other places also. Because of the church's message of justice, I have marched for civil rights, worked in public housing projects, talked to governors and even world leaders, appeared before zoning hearings, written countless letters of concern, and been to countless organizational meetings. I have worked to build a number of institutions to serve the poor, the destitute, and the helpless. They have been good for the people they served.

Then I went to the San Diego Zoo. And I realized that everything I had been taught and had learned, everything I had built and fought for, everything I had given my life for was for naught. Or so it seemed.

Since that day at the zoo, I have been looking at the faith and the church that bears it from a new perspective, from the perspective of Gaia, the living planet, with humanity as one being among many beings. What do we in the church have to say about the creation and the place of humanity in it? How does the faith and practice of the church measure up against the four Dynamics? What does the faith have to say about right relations between humanity and the Earth? These and similar questions have driven me.

This book is the beginning of my answer. Let's explore some of my early thoughts about the place of the church.

Worship

Both in theory and in practice, worship is the central activity of the church. More people attend worship than any other event in the life of the church. So let's start there.

The worship of the church can be inspiring and uplifting. Beautiful music, cogent preaching, graceful and fluid actions all make for marvelous liturgy. But I have become more and more aware that the words and images we use are almost entirely human-centered. Of course, God is the center of attention, but almost entirely as concerned with humanity. Where are the dying species? Where are the polluted waters? Where is the marvel of life? Where is the continuing story of creation?

To be sure, many of the churches in their latest prayer books have added some direct concern about the environment, the creation, God's planet Earth. That reflected a growing concern in the church for this issue. But we still have a long way to go. If we are to respond to this crisis, the central message must expand beyond God and humanity and human activities. The message must reflect that humanity is part of a huge, throbbing life system that is the Earth.

Another central message in our worship focuses on the separation of humanity from God. Humans sin and seek to return to God. That is true. But, from the gestures we employ, the buildings we use, and the words we speak, we seem to understand God as a removed spirit and not much connected to the Earth. We seem to think that humanity's return to God would take it away from the Earth. But it is now becoming harder and harder to understand how we can have that image of a removed God and be rooted in the Earth at the same time.

We must say, of course, that God is transcendent and beyond the Earth. But God is also deeply within us and our experience. God is close, and at the same time beyond and above anything we might imagine. We have much work and imaging to do here if we would have ourselves rooted in God and rooted in the becoming Earth.

Consider how our worship reflects the four Dynamics. First, the Being Dynamic is well served, albeit from the limited perspective of a human-centered creation. The church does, in word and gesture, seek to help people understand what it is to be a human being—humble, loving, honest. But still, that understanding is limited.

Obviously, the sense of belonging is also limited. Humanity does not belong just to itself. Nor just to God. It belongs to the whole creation. So, despite the excellent teaching of the church about belonging—love your neighbor, even your enemy—it is still too narrow an understanding.

Once during a trip to the woodlands of North Carolina, I visited a local church for Sunday Eucharist. As I went up to the altar rail for communion, the fantasy flashed through my mind that the scene would be complete if a raccoon joined us at the rail. Now perhaps I have seen too many Walt Disney movies. But perhaps it was a flash of insight into the presence of the whole creation around the altar in the Eucharist.

Many priests, myself included, have from time to time done a blessing of the animals in our congregations. I can remember parents with amazing good humor hauling all their pets to church on the appointed Sunday. We had cats and dogs, snakes and turtles, even some rather exotic creatures. All were presented to God before the altar in thanksgiving. It was always a special time of belonging.

Now what about the Becoming and Surprise Dynamics? Within the past several decades, we have experienced the becoming nature of liturgy. There has been much change and experimentation in liturgical worship. Prayer books have been revised. New energy has come to liturgical worship. During that period of change, it was pointed out by many that the process was very organic. That reflects the Becoming Dynamic. Liturgy always has a becoming sense to it. The church knows that.

Surprise is more difficult. The church talks a lot about surprise, particularly at Easter. In all too many services of worship, however, everything is astonishingly predictable. Effect follows cause as if the mechanical Newtonian universe is the only reality. Surprise never seems to be a part of these services.

At these services, there is a bulletin with a prescribed order of events. We know what is going to happen next. Even a routine meeting with an agenda has more surprise than these Sunday morning services. Even the prescribed ritual of a parade has more surprise. It is a surprise that more people do not leave in utter boredom. God and the universe are completely tamed and under control.

There is power in the repetition of time-hallowed phrases and actions. The faithful recital of the liturgy gives a sense of permanence in a world filled with human change. That steady repetition even reflects nature's stately procession of days and seasons. The changing of the liturgical seasons can always reflect a similar changing in the natural seasons. The processes of the Earth itself seem liturgical.

The difficulty for many churches seems to be getting surprise into the orderly repetition of the liturgy. How can it be the same and potentially new all at once? Could the surprise of spring be a part of the worship? Could the worshipers join the ebb and flow of the seasons, the newness of the weather, the delight of the bird's song, the awe of the deer's approach? Can there be surprise?

But the actions of the liturgy are not the only important consideration. There are a lot of words in the modern worship service. Despite calls for times of silence in the new liturgies, words often seem to rush along, tumbling over themselves and squeezing out possibilities for any other part of the creation to be present. Some liturgists are quite creative in using silence in a way that lets it speak louder than words. Worship gets closer to the way that the Earth functions when it can creatively use silence.

Humanity is the creation's big brain experiment. And words are the product of our brain. The brain is, of course, part of the creation, and we must listen to it. But the creation is so much more than our brains. In fact, as we discussed earlier, our brains and their words often serve to screen out the rest of creation. So to get into relationship with creation in our worship, we must find ways to get beyond our brain and its words.

It can be instructive to look at the liturgies of the sacraments from the perspective of the larger Earth system. Baptism is seen as entry into the church. But the Native Americans of the Plains, when they welcomed a child into this world, took the child out into nature and introduced that child to the deer and the birds, the trees and the insects, as well as the tribe. For they knew that the child, as he grew older, would have to live intimately with the rest of the creation. So that fact was structured into their initiation ritual. What could we do with baptism to reflect this? For instance, could educated, urban worshipers actually "gather at the river" for a baptism?

What about the Holy Eucharist? The Eucharist is seen as the service of death and rebirth, celebrating the continued presence of Christ in the church. What would the Holy Communion look like if it celebrated surprise in the context of the whole creation? What if the Christ were the Christ who sustains the universe and guides humanity toward its successful speciation within the creation? What would the Eucharist look like then?

What about the setting for worship—the architecture and appointments? In the medieval church, vaulted ceilings reminded the worshipers of the eternal reaches of the heavens. In a typical modern church building, except for the requisite arrangement of cut flowers, we are hardly ever aware of the creation, unless the roof leaks. Light from the outdoors filters through stained glass windows. Stained glass entered church architecture in the Middle Ages to teach visual lessons to the largely illiterate communicants. But that stained glass also betrayed a deep dualism in the thought of the Western church. God was in Heaven and not on the Earth. Heaven was spiritual and good. The Earth was carnal and bad. So the windows not only taught the faith but also had a secondary function of removing any glimpse of Earth while in the church.

Can we any longer afford that kind of one-sided dualism? Doesn't the church need to bring the creation into the church if God is the God of the becoming universe?

Fortunately, more and more church architecture does just this. It has open, clear glass windows so people can look outside and see trees and grass, rain and skies. In some newer churches, light streams in with dramatic effect on the worship space. In the Middle East, many churches are built around rock outcroppings. The tip of the rock will rise up in the middle of the congregation as part of the altar or baptismal font, to be a silent reminder that we are of the Earth.

Some churches now have living trees in them, as do many modern secular buildings. Most churches could easily supplement the requisite fresh-cut flowers on the altar every Sunday with potted plants and even trees in the sanctuary. Some churches now have water running through them as signs of not only baptism but also nature. If we continue to look at our worship and architecture from the point of the creation, who knows what other powerful things we might do?

Other Church Activities

We must direct this scrutiny not only to our worship but also to our practices of Christian education, again usually centered on the human enterprise as if humanity did not belong to a more vast system. Age-related education in the relativistic universe, in modern ecology, in evolutionary theory, and the role of God in each of those areas should be a part of the general educational offering of the parish church. Because of the separation of church and state, the role of God cannot be taught in any of these subjects in our public schools. Because of right-wing religious pressure, some of these *topics* cannot be taught in some public schools even as a secular subject.

In addition, specific education in the particular ecosystem in which the parish is located would be of great benefit. What species are present? Which are endangered? How do they relate to each other? How is the presence of humanity making an impact on this particular ecosystem? Where can the church people see that being played out?

Finally, education in sustainable living should be offered. What practice beyond the recycling of old newspapers and plastic will be a contribution to the ongoing vitality of the Earth? What sound construction methods can be safely used? What about transportation, lifestyle, the use of money and investment, recreation? What can the members of the church do to have fun and be kind to the Earth and its species at the same time?

Another activity, pastoral care, concerns the life of the community and its individual members. As pollution grows worse and the loss of species develops, there will be increasing anxiety over the future. This must be addressed not only through education but also through pastoral care. During the use of each of the pastoral offices (confirmation, marriage, burial, etc.) the accompanying instruction should address, among other things, the larger Earth community in which we live. Each of the offices should be tied in with the larger life story of the universe. As we discussed earlier, the fear of death drives much of the ecological crisis. Death must be restored to its proper place in the human life process. Death is real enough, but is not to be feared so much as welcomed as an important contribution to God's creation.

Outreach, or community ministry, concerns itself with the outside world and people beyond the church. It is essential that the church link its activities together with other community groups working in the area of the environment, ecology, and sustainable development. The church needs to be at the table making its contribution. Those efforts should address public policy, private activities, and the common good of the Earth. It might even be possible for larger congregations to sponsor the development of sustainable communities, building and operating them for their own members and others.

Stewardship and finances are also important in the life of the church. Good stewardship teaching and practices will not just pose humanity as responsible for God's creation, but will firmly set humans as one species among many, living together on the Earth. The church's finances will similarly reflect creation sense. Any investments, for instance, should be made in companies that have good environmental policies. There are a growing number of investment counselors and mutual funds that point to good environmental investing.

The church has responded to new contingencies, new concerns, many times in the past. Despite or maybe because of its roots in history, the church can be quite flexible. That flexibility attracted me in the beginning and keeps me attached. Movements have swept through the church and stirred up incredible creativity and passioned response. It is time for such a movement now.

Theology

Behind life in the church lies theology. We live out in the church our understanding of God and the universe. At the present, our theology is not yet adequate for a universe of becoming and surprise, being and belonging.

First, our theology is human-centered. Despite the fact that God came in Jesus to save humanity from itself, humanity is not the only reality in the creation. Humankind is one species among millions on the Earth. The Earth is one planet circling one star among billions in the universe. Each human individual belongs to a species of six billion people. And that species belongs to a network of life that encircles the globe. To concentrate our theological inquiry on each individual human to the exclusion of the larger belonging is inadequate at best.

Modern theology talks almost exclusively about human-human and human-divine relationships. But what of the human relationship to the rest of organic life and to the rest of the creation? It is one thing to speak of the salvation of individual humans. But what of the preservation or salvation of the human race as we careen toward ecological disaster? And what about the salvation of other endangered species? Modern theology is almost completely silent on these subjects, though there are encouraging stirrings, which are quite important.

Theology reflects our culture. In our culture, the solitary human is the center of concern, both for individual action and for relationship. We are the "Me Generation." We operate in a free enterprise economy. The question often posed by our culture and economy is, "What shall I, as an individual, do?"

As has been widely observed, that is a limited question. It does not reflect the whole of reality, which is far greater than one individual or even human society as a whole. Evolutionary thought suggests that species, not individual organisms, are the primary actors on the stage of history. To be sure, early in the speciation of any species, individual organisms make crucial decisions. That is where the human species is—at the point of crucial individual decisions about the future of the whole species. So we can't eliminate the theology that talks about the individual.

Theology now, however, also needs to address the whole. We need some thinking about systems and the relationship of God to those systems. Theologically, how does the whole belonging work? What is the relationship of each being within the system to the whole system? What do the incarnation, life, death, and resurrection of Jesus have to do with that whole system? How does God relate to that system?

For the first time in centuries, we now have a scientific cosmology that is fairly well accepted around the world. Einstein's thought is generally agreed on. We are in a relative universe where energy and matter are two sides of a single coin. There is no dominating dualism in our modern scientific conception of the creation. All is one. But theology has said little about Einstein.

The urgent task for theology is to describe this Einsteinian cosmology from the point of view of the Christian revelation. It is a task similar to that which theology always faces. How do Christians in any age understand the prevailing cosmology in light of the revelation of God in Christ?

Some branches of modern theology have much to contribute. Process theology is built on the understanding of a dynamic and evolving universe. What do modern process theologians say?

Feminist theology looks at the world from the unique perspective of women and the way they relate to life. It is a rich, emerging discipline. Feminist thinkers say that the issue for women is the same as the issue for the Earth: domination and exploitation. So freedom, equal rights, and nurturing are essential to the survival of both women and nature. What else might feminist theology say about these issues?

Liberation theology looks at life through the perspective of subjugated peoples. Since the Earth is being subjugated by humanity, liberation theology could have an important contribution to make. All of these new theologies must address new understandings of the relationship of humanity to the Earth.

Some recent creation theology uses stewardship thinking to get at the issue. The basic premise is that humans are to be stewards of the creation. God owns, and humanity runs, the creation. Actually, contrary to stewardship thinking, the creation takes care of humanity more than humanity takes care of the creation. Humans must find a way to see themselves as one, albeit quite talented, species among a host of other species, and those among a complex Gaia system. Stewardship thinking just doesn't call humanity to a radical enough repentance or a deep enough humility. Only when humanity understands that it has not been granted any control or dominion or special dispensations can it start to regenerate a sustainable place for itself on the Earth. Only then can humanity move beyond the Love Canals in which it has become immersed.

The theological thinking necessary for survival is just starting. Much work needs to be done, both at the level of formal, academic theology, and at the level of theology that is preached and lived, the theology that undergirds the decisions we make every day of our lives.

The Task for the Church

Finally, having looked at what the church does and what it thinks, we ask what the church can contribute to the becoming universe. What can any of us do? I am, personally, just a small part of the church, which is a small part of the human race, itself a small part of the Earth. But I wake up early many mornings, dreaming of the church and of the humanity that must be. I have thought and prayed, listened and wondered, talked and written, searching for my place, the church's place, humanity's place on this Earth. So far, I have more questions than answers. But they seem to be the right questions. And that is crucial.

Whereas my contribution right now is mainly thinking, other members of the church are also active, working on personal and congregational lifestyle changes, political changes, educational changes, and so on. A growing ferment of activity seeks to respond in faith to the becoming universe. Progress proceeds too slowly in many areas but offers inspiration in others. The need for more people to become involved in doing what they can do is urgent.

The church has a long heritage rooted in the life of the people, connected to the culture, and open to new possibilities. Its Scriptures are rich in wisdom, its theology wise in the ways of God and the people. The church is positioned to make an important contribution to the future of humanity. Can it respond?

Finally, above and beyond our membership in the church, we do all that we do as participants in a wonderful universe—heirs of a series of events that have brought us to where we are, and seekers after the next series of events, which will bring us to where we will be. The decisions we make are crucial and, at the same time, will make no difference. For our task is to give up control, turn ourselves over to the becoming universe, and work diligently with that becoming universe. Because finally, it is God's story, not ours. In that we can rejoice!

CHAPTER 6

And There We Go

THE EARTH AND its creatures are calling out to humanity to take its rightful place in the creation. It is a subtle but desperate plea, with little apparent response on humanity's part, to this point. Take that day in the mountains . . .

Ed and I set out early in the morning to drive to the Cohutta Wilderness of North Georgia for a hike to the Jacks River Falls. Parking at the trailhead rather late in the morning because there had been so much to see just getting there, we headed off down the trail. In no time, we were engulfed by the forest, caressed by its cools and warms, lights and shadows, and sung to by the little brook that hugged our path, crossing and recrossing it at times.

We hurried on because we wanted to get to the falls, a picture of which we had seen in a promotional folder. After a couple of strenuous hours, we came to the Jacks River below the falls—wide, deep up to our thighs, and cold. Despite a brief thought of declaring the river to be the end of our hike, we plunged on, getting a bit wet and emerging on the other bank exhilarated and ready for the prize of the falls.

Fifteen minutes later, the trail rising high above the river below, we came upon the thundering falls. After an exchange of triumphant high fives, we fell into an awed silence before the scene. A torrent of water plunged down the gorge, sending a mist into the air and

131

watering the lush foliage all around. The air filled with the roar of waters so that we had to lift our voices to be heard by one another.

Slowly, we picked our way down onto the massive rocks by the top of the falls and sat down in reverence. We could almost feel the rocks quivering under us from the pounding flood.

Ed, given to ritual, was making one of his offerings to the water when he turned to me and said, "Look up." There circling over us were 10 or 12 large birds. "Those are eagles," Ed explained.

I looked more closely. Different from the hawks or buzzards that you generally see, they weren't just circling, waiting for the death of some creature below. They were weaving in and around each other, as if in some playful dance.

We watched for some moments, the waters booming, the ground shaking, the eagles frolicking, and Ed and I entranced at the wonder of it all.

A holy moment.

Presently, Ed stirred. He had some small binoculars and wanted to take a closer look. As he raised them to his eyes, the eagles all turned and, as one, flew away from us, the communion broken. In a few moments, Ed and I turned to leave also.

Later, a friend of mine observed that the eagles had been there to greet us but had been frightened by the glasses. "They can," he reminded me, "see a field mouse from a mile away. They surely could see what you were doing and didn't trust your intentions."

For a moment we had been one with those eagles, in as much communion with them as we could be with any creature. For a moment there, we had the experience of right relations between us humans and the creation around us. It was deeply satisfying.

WHEN ED AND I got back to the city, life was going on as usual, marked by confusion, suspicion, exploitation, and the signs of the end of the era of humanity.

What a marked contrast with the peace of the Cohutta Wilderness. It is increasingly difficult to live in this human-dominated world, particularly against a remembrance of occasions like that day with the eagles.

The historian of cultures, Thomas Berry, says that we are at the end of the Cenozoic era. Just like the great extinction of the dinosaurs and the emergence of mammals marked the end of the Mesozoic era 65 million years ago, so the extinctions going on all around us now mark the end of this era. We are in the middle of a great crash of life and the emergence of something new. The new era dawning, if we survive into it, will bring with it a new sense of the place of humanity on the Earth. With it will come a new cosmology.

Berry suggests that the next era will be the Ecozoic era, meaning that surviving humanity will have to learn to live ecologically in that new time. What will that time look like? How will we humans live in that era? What will be our cosmology? What instructions do the four Dynamics give us?

THE ECOZOIC ERA

The growing environmental crisis is the means by which the Earth is seeking to get the attention of humanity, telling it to come to its senses if it would survive. It is like the legendary farmer hitting the mule on its head with a 2×4 in order to get its attention. The Earth seems to be hitting humanity on the head so that humanity might change.

But the message does not easily get through. Recently, I stood at the curb during a fall college homecoming parade. Next to me, a young couple was talking about the unseasonably warm weather. One remarked that he remembered when he had to really bundle up for this parade. The other responded that she didn't know why things were so different.

Now maybe that warm weather had nothing to do with global warming and the ecological crisis. But in that particular place, according to the newspapers, that month was the 22nd straight month of above-average temperatures. But it is hard for us to put all of that together. It is even harder for us to imagine how we will have to change.

The Being Dynamic in the Ecozoic Era

The humanity that emerges from this crisis will have to live by each of the Dynamics. In its being, it will be much more humble, realizing the natural limits on its wisdom and power. Humanity, despite its vast knowledge, is not wise enough to remake the creation. The arrogance that tempts humans to remold the face of the Earth for the sake of human "progress" will be crushed under the relentless response of the Earth, driving humanity to its knees in humility before the wisdom of the creation.

The big brain of us humans has lured us into pretensions about our abilities that will not stand the test of time. We ravage the Earth with the powers inherent in this big brain. There is no way not to have a brain. We cannot all have a prefrontal lobotomy. To what ends must our brain be given?

For the first time in all of creation, the Earth has produced in the human brain a way to think about itself. The self-consciousness of humanity provides the possibility for the Earth to be self-reflective. We humans are the Earth, thinking. Until this time in history we have been preoccupied with our own individual or tribal fortunes. The negative consequences of that preoccupation were not apparent. But now we see that this preoccupation is actually endangering us more than it protects us.

The present times teach us that human thinking cannot be just for the sake of humanity. It must be for the sake of the Earth as a whole. Now, through human beings, the Earth has a chance to consider itself and its future. It has a chance to direct its own evolution, ending it if it would, or changing it in more positive directions. But the thinking must be for the whole. No part of the whole can be isolated from the whole.

The power of humanity is enormous. We have seen what it can do. Now, even while we are thinking for the whole, we must curtail the use of our power. Humanity must develop restraint. We must grow up to be a gentle giant. Since we know our capacity to destroy, we now must discover our capacity to preserve. The new being of humanity must be marked by its capacity to save, guard, conserve, and rescue. Instead of conquerors, we must now be guardians.

We must also think in terms of the right scale. Cheap fossil fuel energy has seduced us into a scale way beyond what is appropriate for sustained life on the Earth. The Earth has a deep wisdom about how to proceed according to scale. Things never get bigger or more flamboyant than that which can be sustained. Or if they do for a season, they are eventually called back to size. In the Earth system, larger is not better.

Humanity will live more appropriately than it is currently living. Dwellings will be smaller, buildings scaled to human proportions and the shape of the land, businesses more local and closer to the Earth. Air will circulate more freely and water will be used more prudently. Vegetation appropriate to the climate and closer to the natural will be used. Every lawn does not have to look like a putting green, and every house does not need a swimming pool and a whirlpool.

Sufficiency will mark the new being of humanity. More will not be better, but enough will be sufficient. There is a limited amount of food, energy, and material resources in the world. The new human will have a deep sense of minimalism, living on just enough for sustainable living. The world is not infinite, and humans are not just unbridled appetites.

Humans are called primarily to be producers and creators, not merely consumers and takers. Our economy must prize elegant simplicity, not opulent consumption. This style offers ample opportunity for creative expression in art and politics, literature and discovery. But it will no longer venture down the illusive path of indefinite expansion. The Earth will not sustain that.

Walking through an art gallery, I am always struck by the early-nineteenth-century painters, John Constable (1776–1837) and William Turner (1775–1851). They specialized in landscapes and depicted the energy and strength of nature with vivid emotion. In their paintings are huge rock formations, wildly tumbling waters, towering clouds, and flourishing trees. Humans were in their paintings, but always as a small part of the vast canopy of nature. So it must be for humans in the Ecozoic era.

In summary, the sense of being in the new human will allow that human to think for the whole but at the same time be marked by an incredible modesty. There will be a strong sense of humility before the wisdom of the creation. In its cosmology, the new humanity will understand itself as far from central in the workings of the Earth. In its actions, the new human will be quite creative and will build to scale in the midst of "this fragile Earth, our island home."

The Belonging Dynamic in the Ecozoic Era

The Belonging Dynamic must also be applied to the new human. As we have pointed out earlier, the Earth is one huge living system— Gaia. It pulses and breathes, acts and reacts. Humanity is a product of the Earth and is absolutely dependent on the Earth. We humans belong to the Earth.

This Earth system can be thought of as an immense community, a community of subjects. Despite human pretension, we are not the only subjects in a world of objects. The other beings in this system are not just objects, resources for us to use as we see fit. All other beings are also subjects with identities and roles.

If they are subjects, they must be related to as subjects. Martin Buber (1878–1965), the great Jewish philosopher, talked of I-Thou relationships as the only way for subjects to relate.[18] If we relate to another subject in an I-It relationship, we devalue that subject and devalue ourselves at the same time.

Not only must we relate to the rest of creation as subjects, but we must value the great diversity of beings on the Earth. Diversity produces the difference needed for sustainability. Lately, we have been building a global human culture and eradicating human diversity in the process. Blue jeans and rock music are found all over the world. We have put all of our eggs in the same basket, so to speak. And that basket is increasingly Western high-tech society. This leaves us dangerously vulnerable to a fatal disease, or the end of sufficient energy resources, or the dictates of an overheated planet.

Diversity in human populations can exist only within a world of respect and appreciation for other human cultures and ways. It works the same with nature. The web of life is built on a complex pyramid of diverse beings. No one knows when the decrease in diversity will lead to a great collapse of the system.

I'll never forget sitting under the vaulted ceilings of the National Cathedral in Washington, D.C., listening to William Reilly, the director of the U.S. Environmental Protection Agency. Reilly was talking about the loss of species when he looked up at the towering

---

[18]*I and Thou* (New York: Charles Scribner's Sons, 1958).

columns around him holding up the arching ceiling. He remarked that he could start pulling out stones, one by one, from the columns and not much would happen at first. He could continue to pull them out, and still nothing might happen. Then, at some point, he would pull a stone and the crash would start and the whole edifice would collapse. That is what we are doing with the creation.

Diversity marks the resiliency of the Earth system. Diversity builds in a margin of safety from assaults on the system. Diversity brings with it stability. Right now, we seem to abhor diversity, in the human community as well as in the Earth community. The new human will cherish diversity.

This Earth community must also live together in a balance of sorts. Justice must prevail among the subjects within the Earth community.

The category of justice relates foremost to humanity. Economic justice for humans is inextricably linked to environmental survival, as we saw from the early report from Gro Harlem Brundtland and the later report from the United Nations, GEO-2000. For humanity to survive, there must be economic justice among humans. Those who are starving must have adequate standards of living if we would save the environment. They must not be forced to destroy the Earth in order to survive. It is a matter of justice. It is a matter of belonging.

In addition, Brundtland's Commission pointed out that improvements in living standards have often been achieved through the creation of pollution or a ravaged Earth not adequately paid for through the price of the improvement. Clean air and water, replanted forests, and restored topsoil must be paid for in the cost of the goods themselves. Some of us have been growing rich not only at the expense of each other, but also at the expense of the Earth itself.

But the same exploitation is true among the various species of the world. The stronger species such as humans are destroying the weak species. The rights of those more vulnerable species are being trampled. Isn't that also a question of justice?

For example, in the Pacific Northwest of the United States, the "salmon wars" are heating up. Before all of the dams were built in the Columbia River, 10 to 16 million salmon made their annual run up to their spawning grounds. Now barely a million do so annually and it is getting worse. In the meantime, dams were built across the river, which, despite their "fish ladders," effectively block salmon from their migration. Houses and office buildings were built in wetlands, depriving the salmon of spawning grounds. Lawns are fertilized but the fertilizer runs off and destroys the streams. The population of automobiles explodes and their emissions into the air and the runoff of gas and oil into the streams further endanger the salmon.

If the salmon are to survive, it will take a huge change in lifestyle for the humans in the Pacific Northwest. Will they be willing to do so? For the sake of justice?

Justice for all species will characterize the world in the Ecozoic era. There will be respect for the place and contribution of each, and a desire of the whole to preserve the constituent parts. Now that humanity can think for the whole, there is no avoiding the responsibility of humans to establish justice for the whole.

Couldn't we, for instance, establish a "Bill of Rights" for all of the species of the world? They too have the right to "life, liberty, and the pursuit of happiness." Why shouldn't this highest achievement of the human race also be applied to the rest of creation? In English law, the rights of the weakest are guaranteed in order to protect the rights of all citizens. The framers of this code of justice rightly understood that the welfare of the strongest is dependent on the welfare of the weakest. Aren't we now seeing that this is also true of all beings in the creation? The survival of the strongest absolutely relates to the survival of the weakest.

Why can't courts protect the rights of the different species? Humanity, with its ability to think for the whole, would have to run the court. Omsbudsers would speak for the other species. Grievances involving dangerous activities by any species, particularly humankind, would be brought before the bench and tried. It would be a point of justice.

We are finally talking about right relationships. In the belonging universe, humanity would be in right relationship with the rest of the creation. If the human species is to survive, those relationships, which have been destroyed by human beings, must now be set straight. Everything belongs to everything else on the Earth. Soil belongs to worms, which belong to birds, which belong to trees, which belong to air, which belongs to humans, which belong to soil. All belong.

I have participated in a conference experience called the Council of All Beings, which pushes toward this understanding.[19] Before a typical council starts, the different human participants let themselves be "chosen" by another being to attend the council. One might be a worm, one a mountain, another an owl, and another a river. Each participant then makes a mask to be worn when speaking for the represented being. All speech occurs in the first person.

At the council, the assembled beings take turns addressing human beings out of their pain, their grief, their joy, their bewilderment. The speeches are often passionate, are sometimes funny, and are always gripping as the stories of the assembled beings are laid at the feet of humanity. Each participant also has a time during the council to be addressed as a human.

Finally, after all the stories are told, the humans respond. They are penitent about the destruction they have caused. They ask for the gifts that the various species bring—the wisdom of the owl, the industry of the ant or the beaver, the tenacity of the mountain, the life of the mountain brook. As the gifts are offered, the participants join the rank of humans in the center, their eyes opened to the incredible community of belonging in which they live. People leave the council with a different understanding from when they entered. It is very much like that day with the eagles.

So the new humanity required for the survival of the species will need a modest sense of being and a strong sense of belonging. Thinking back to the diagrams for the relationship of humanity to the creation that we have been developing, you will remember that the present understanding of Western culture might be diagrammed like this:

Domination model

Humanity

God

The creation

---

[19]John Seed and others, *Thinking Like a Mountain* (Philadelphia: New Society Publishers, 1988).

We do see nature in our day as a very complex belonging, so it is represented here as such. But humanity is still outside of and dominant over that belonging.

We also referred earlier to a religious model that had the primary relationship between God and humanity and had the creation outside in a secondary role.

In the Ecozoic era, the model will be all-inclusive. It might be called the Ecozoic model and be diagrammed like this.

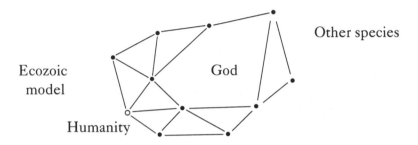

God is in the center of the creation and the being of humanity has joined the belonging of the creation. Humanity is depicted as a large being because of its power. It is positioned near the periphery as a reminder that it is a new participant in the large, complex, diverse community of many different species, both plant and animal.

It is like living in any neighborhood. You move in and then, after a while, you start to wonder about some of the neighbors. Some are quite strange. Others are actually hostile and dangerous. Some of those neighbors probably look at you the same way. But you have to learn to live with them. You have to respect their rights to live there. Otherwise, the neighborhood will go up in the smoke of internecine warfare, where all will lose.

To push this Ecozoic model a bit further, how do we live with the dominant strength of the human species in this community? How can the dominant human species coexist with weaker species?

For a clue, we might look at the place of the United States in the world community. The United States is far-and-away the dominant national power in the world. Economically, militarily, culturally—the United States is the strongest.

So what does it do about the smaller nations, particularly the troublesome ones? They cannot be just wiped out. Nor can they be ignored. The United States must live with them, even help them to survive. The survival of the United States itself is at stake.

Through these models, we push toward a new cosmology. No longer is humanity outside of the creation with the creation operating mechanically, like a giant clock. Now humanity is an integral part of a large, complex system that is quite organic in its workings. The relativity of Einstein, the unpredictability of quantum mechanics, even the turbulence of chaos theory are all part of this emerging cosmology. Perhaps it will be a cosmology to survive on.

Now let us consider the Becoming Dynamic. Where will that lead us?

The Becoming Dynamic in the Ecozoic Era

The Becoming Dynamic points to the dynamic nature of the universe. It sees the universe as an unfolding expression of God, begun at the Big Bang beginning but continuing toward greater and greater complexity and more and more consciousness up to the present and into the future.

In terms of complexity, humankind is surely the most intricate, complex, and able expression of this creation that we know. What a wonder! With our big brain, we dominate the Earth, even to the point of destroying it. We sing and dance and create in ways that no other species has even dreamed of.

It is tempting, therefore, to think that humanity points the direction in which the creation is unfolding. It is even tempting to think that the spirit world, the world to which we relate by our "sixth sense," the world coming after the four dimensions of space and time, is the next stage of evolution.

In this projection, our physical selves are less and less important. The life of the spirit becomes critical. Science fiction is one way to speculate about this possibility. Some science fiction writers even see humans of the future as disembodied spirits.

I have been troubled by this projection. It seems to fit right into the one-sided dualism that has plagued our thinking. Does this understanding of evolving to the spirit world not say that the spirit is superior to the material world? Does it not tempt us to escape from the material world? Does it not give us implicit license to devalue, and then plunder, the material world?

So before we seal this conclusion of "progress" toward the spirit, let us look again at the Becoming Dynamic. Is the universe becoming spiritual? Is the universe moving from the physical toward the spiritual?

Could it be, rather, that the universe is moving from the spiritual to the physical? Could it be that the energy-particle composition of the earliest universe was grounded first in the spiritual dimension and then participated in the next four dimensions of space-time?

From this point of view, the first dimension in the creation is the spirit. The dimensions of space-time are then added to the spirit dimension. "In the beginning was the Word" (John 1:1). In the beginning was spirit. It was the first reality.

As we consider this idea, we must again remember that God moves in the spiritual no more than in the physical realm. God is the reality upon which both rest. We do know that our physical senses, operating in the dimensions of space-time, are unable to access the spiritual. But what if the universe started with the spiritual? What if it were only spirit for an instant at the beginning of the Big Bang? "Let there be the dimension of the spirit." And it was so.

Immediately afterward, the next concrete expression of spirit—the space-time dimensions—expressed itself as exploding, swirling energy that condensed into elementary particles and then evaporated back into energy. Gradually, the creation became more complex and conscious, as we have seen. Finally, biological consciousness came and then, in humanity, the big brain with its self-consciousness. But at the base of the whole story is the dimension of the spirit. It is not the higher dimension, but the most fundamental dimension. The whole creation is built on spirit and is a great flowering of the spirit.

The latest product of this spiritual-physical-temporal world is the human mind. Like a kid with a new toy we spend our days playing with this mind—thinking, scheming, fearing, protecting. It is a powerful toy. But our love affair with the mind has blinded us to the deep sea-currents of the creation, where the Spirit lives and moves and has its being.

I wonder if the spirit realm is the realm of instinct and intuition. From this point of view, when a monarch butterfly migrates from Canada to Mexico and takes three generations to do it during the course of a solar year, it flies by the spirit. Now, to be sure, those migratory routes are coded into the DNA of the butterfly. There, the instinct-DNA codes help the butterfly steer according to the sun, the weather, magnetic fields, and so forth.

Those instinct-codes are stuff fundamental to the planet—spirit stuff. When a person acts on a hunch that is beyond rational or sensory explanation, that person participates in the spirit. The spirit is giving deep, Earth wisdom—God wisdom—to the person. Without the guidance of the spirit world, our noblest schemes are therefore doomed. We wander aimlessly with no meaning or direction. With our great reliance on rational mind and our distrust of intuition, no wonder the environment is so endangered!

I've wondered if the spirit might be the deep, primitive movement of the creation. The spirit is the first expression of the personality of God, the life force that drives all that is. The spirit is the first-created, the Christ of the four Dynamics. It is fundamental to reality.

It is then understandable how many Native Americans saw their spirituality. Their cosmology had physical reality based on spirit. Spirit was not beyond the physical. It was the first dimension and the undergirding of the physical world.

With spirit at the base of the creation, there is no possibility for a crippling dualism. Spirit is not higher and beyond the physical. It is the rock upon which the physical is constructed. This does not mean that humanity can no longer inquire into the ways that the physical world works or that it cannot have technology or do thinking. It does mean that our science and technology must be guided by the deep spiritual wisdom of the creation.

Humans have much more difficult access to the spirit than do animals. Our big brain has seduced us into brain-reliance rather than reliance on the basic currents of the creation, as was the case for countless generations before us. Enthralled by our minds, we miss the subsensory beat of the creation. In terms of the diagram that we used to show self-consciousness, it is as if we see the mind as reality, rather than acknowledging that the life of the spirit as expressed through our bodies is the reality.

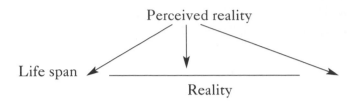

Reality is primarily in the spirit and in the body, not in the mind reflecting on the spirit and body. That is why the great mystics who understand the world of the spirit spend long contemplative periods in silence. They clear the mind and listen with all their senses, including their sixth sense. That is why Jesus, when he wanted to pray, left the world of humans and went off into nature on a mountaintop or a wilderness or a garden. He went to get away from the babble of mind.

The mind is called to think about and reflect on what the body is experiencing. The mind needs to work with the body, not replace it. We are talking about "mind with matter," not "mind over matter."

From this point of view, the becoming universe is a flowering of the spirit that gives life and meaning to the whole creation. The physical world, the world accessible by the senses, is built on the spirit and is, therefore, derivative of the spirit.

So what is the creation becoming? It is becoming what it has always been becoming: a vast, fascinating display of the glory of God, increasing in complexity, in consciousness, and in diversity.

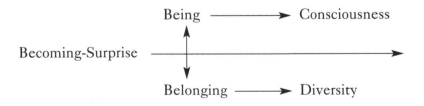

What is humanity becoming? We have choices. We can become the thinking element of the Earth that will, for the first time, think about the Earth as a whole. Or we can become nothing. The brain must now integrate with, and serve, the Earth. If we don't get our big brain under control for the sake of the whole, rather than letting the big brain seek to control everything, we will become nothing.

And how will humanity move on? That is a matter for surprise.

The Surprise Dynamic in the Ecozoic Era

The Surprise Dynamic, seen most clearly in the resurrection of Christ, describes the method by which the universe is becoming. From the human perspective, what happens in the creation is so often unexpected. It is a surprise. From the universe's perspective, it is a process of death and life, life and death. New life always seems to emerge out of the demise of an earlier form. A species will go extinct, leaving an empty ecological niche. Into that vacuum will evolve new, more sophisticated species.

Humanity must participate in surprise if it would grow to maturity, if it would become an established species, living in harmony in its ecological niche. So what does that mean?

We humans must give up our attempts to control nature and to control even our own destiny. We must give up control and let the deep currents of the creation lead us to life and safety. This doesn't mean that we must become passive and drop any initiative. Rather, it means that we must work as hard as we can and then know when to turn our efforts over to the greater wisdom. It is as simple, and as complicated, as that.

The new cosmology, built on Einstein, quantum mechanics, and chaos theory, gives no room for control. So how does an entire species give up control and trust the creation? Well, it must start with some individual members of the species, realizing that the species is in peril. Some individuals must take it on, one by one and in small groups, and lead the rest of the species into the completion of our speciation. Every species that has eventually found its niche first went through a time of experimentation and trial, as individual organisms tried an adaptation to see if it "worked." Some did and some didn't. But if enough tried the same thing and those attempts fit the niche, then the species could stabilize and survive.

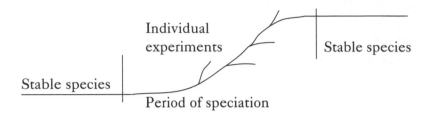

Jesus was one of those individual organisms that made the adaptation. Other religious leaders—and some, perhaps, not so religious, such as Socrates and Aristotle—have also been the experiments of the human race to find the complete speciation of the race. It is crucial that still others try.

It is also possible for experiments to be undertaken collectively. Some utopian groups have tried this. Even whole countries such as Costa Rica have made attempts to live out the new model of harmony between humanity and the creation.

Consciousness is the very ability that we now need to complete our speciation. We have talked earlier about the universe becoming more conscious until it reached self-consciousness in humanity. The universe now needs to become even more conscious. It will do so through humanity.

WHOLE-CONSCIOUSNESS

The step beyond self-consciousness is whole-consciousness. Humanity must be conscious not just of itself, but of the whole. We are the Earth, thinking. We are the Earth, conscious. We are the Earth, able to choose the direction of its evolution. We must have a consciousness of the whole system, if we would do that.

The primary reality is the whole. Despite our pretensions, we humans are not the primary reality. The question before us is where we fit into the whole, not where the whole fits into us. That is a monumental shift that the human race must make. It must make that shift to survive, just as those early trilobites and every other established species did.

Peter Russell, the British psychologist, entertains a fascinating theory in his book, *The Global Brain*.[20] Summarizing contemporary systems theory (p. 56), he points out that the complexity of any system can be measured in terms of three basic characteristics. (1) Quantity/diversity. The complex system contains a large number of different elements. (2) Organization. The many components are organized into various interrelated structures. (3) Connectivity. The various components are connected through physical links, energy interchanges, or some form of communication.

In terms of quantity/diversity, it appears that a new evolutionary level emerges in the universe when the quantity of organized organisms grows to about 10 billion. When about 10 billion atoms gathered together, they could organize themselves into a simple living cell. Few living cells have less than that.

There are about 10 billion nerve cells in the cortex of the human brain, the critical mass that led to the emergence of self-consciousness. Smaller brains just can't be self-conscious.

Likewise, each of these levels of increasingly complex life is increasingly organized and has increasingly sophisticated means of communication. The human body is at the pinnacle of complexity and is a marvel of organized organs and systems that communicate with each other through a complex nervous system.

Russell observes that the Earth is approaching a point where the population of human beings will be 10 billion. Human society is getting increasingly organized. Many political and economic organizations have worldwide influence. Worldwide communication links are being put in place at breakneck speed through the Internet, telecommunication, and travel.

With this stage set, there is the possibility for something new. Russell speculates that the self-conscious humans of the world will get so organized and develop such communication links that they will start to function as one giant brain—a global brain.

Humanity would start to think as one. There would be an organic whole-consciousness. It would be like Jung's collective unconscious,

---

[20]Los Angeles: J. P. Tarcher, 1983.

except it would be a collective consciousness. It would be a leap like none that the Earth has previously known.

In that new evolutionary level, each human would have a definite purpose, just as each cell in the human body now has a definite purpose in relation to the whole body. Humans would be organized into functional units just as we are currently organized into families, businesses, and associations. It would be analogous to the way that cells are organized in the body into organs. But the units would work together in harmony rather than in random opposition, as humanity currently seems to operate.

Does the cell cease to exist because it serves a larger body? It is the opposite. The cell flourishes as it serves its role in a larger, well functioning body. We even say that healthy cells "glow" in a healthy body. So it will be for the larger planet. Individual humans will glow as the larger body of the Earth flourishes. It would be like a championship basketball team. As the team works together more harmoniously, the members develop into better players individually. Just look at the 2000–2001 edition of the Los Angeles Lakers.

What is needed is an increasing number of individual brains that have developed whole-consciousness. In fact, it is imperative for the future of much of life, including human life, for an increasing number of humans to become conscious beyond their narrow self-consciousness. It is critical that more and more individuals move into whole-consciousness.

Now Russell's theory of the global brain is admittedly speculative. We have no way of knowing if the universe will develop in that direction. The future direction will be a surprise, anyway. If we knew what would happen, it wouldn't be a surprise. The future direction of individual humans must be in the direction of whole-consciousness.

God has showed us in Jesus, the Christ, what we must do to have whole-consciousness. We have discussed that course as we looked at the great Dynamics of Being, Belonging, and Becoming. Finally, there is surprise: taking initiative, giving up, and letting go. That is where Christianity has a vital contribution to make. Let us look more carefully at the way that Jesus points.

THE WAY OF THE CROSS

In theology, we say that Jesus walked the Way of the Cross. This was the way to Golgotha, the place of the crucifixion. The Way of the Cross, as we think of it theologically, is a focused recapitulation of all of Jesus' teachings and actions. It is the way characterized by a connecting love that risks vulnerability. There are no self-protections, only love. It is the way that is the opposite of control.

Jesus walked the Way of the Cross and offered himself in humility on a donkey in the entry into Jerusalem. He cleansed the temple of the moneychangers who would exploit the creation for personal gain. He championed the downtrodden, the weak, and the insignificant. For their sakes, he put himself at risk again and again.

All of this threatened the authorities and increased the danger that he would die. He was frightened and wrestled in himself about whether or not to continue. He decided to press on and was arrested. Again, he was tempted to slow his pace, protect himself, and try to control the situation for his own personal survival. He was tempted not to care for the whole. Again, he decided against that course and proceeded in trust.

Finally, the end came. He was crucified and died. It seemed over. The disciples feared for their own lives and scattered. The lesson seemed to be that unless you protected yourself, there would be no life, but only death.

Then the surprise came. In a mighty act of becoming, God raised Jesus from the dead to new life. And the affairs of humanity have not been the same since. This is the Way of the Cross. Each of the Gospel writers describes it as a central theme.

"Then he began to teach them that the Son of Man must undergo great suffering, and be rejected by the elders, the chief priests and the scribes, and be killed, and after three days rise again. He said all this quite openly" (Mark 8:31–32a; see also Mark 9:31; 10:33–34; Matt. 16:21; 17:22; 20:18–19; Luke 9:22; 18:32–33).

St. Paul sets the Way of the Cross into the whole incarnation and describes it in soaring poetry, bidding Christians to follow this same way:

Let the same mind be in you that was in Christ Jesus,
    who, though he was in the form of God,
    did not regard equality with God
    as something to be exploited,
but emptied himself,
    taking the form of a slave,
    being born in human likeness.
And being found in human form,
    he humbled himself
    and became obedient to the point of death—
    even death on a cross.
Therefore, God has highly exalted him
    and gave him the name
    that is above every name,
so that at the name of Jesus
    every knee should bend,
    in heaven and on Earth and under the Earth,
    and every tongue should confess
    that Jesus Christ is Lord,
    to the glory of God the Father.   (Phil. 2:5–11)

The resurrection was the life-giving surprise. Three days after the crucifixion, Jesus appeared first to the women, and then to some other disciples in his small band. He ate with them and talked with them. Yet, he was not the same as before. He appeared to them suddenly and mysteriously, even though they were behind locked doors. He was with them, though unrecognized, and then was suddenly recognized through a spoken word or the breaking of bread. Thomas wouldn't believe that it was he until after he had touched his wounds. Jesus was real and unreal at the same time.

So what is to be made of the resurrection? Is it just a fantasy on the part of the disciples? Is it just proof of an afterlife? Or is it more? Could it even be more in terms of the whole creation? Theology has much to discuss about the connection of resurrection and the life of humanity in the whole creation. Consider the following possibilities as places for theology to begin.

The resurrection confirms that Jesus was the human expression of the ultimate life force of the universe. Jesus was the temporal appearance of God, whose creative energy had burst forth at the creation, manifested in hydrogen atoms, appeared as magnificent galaxies, emerged as primitive self-replicating life in the ancient seas, grown into more and more complex and conscious life until it had finally come in the form of a human—Jesus. The resurrection was the surprise exclamation that being, belonging, and becoming were the story of the universe. They could not be squelched or killed. They are forever. And they are in clear focus in Jesus, the Christ.

The resurrection is also the statement that the life force of the universe will go on. Jesus was not the end. Humanity is not the end. The universe is still becoming. Surprise is still the operative mode of progress. Humanity is an expression of this deep, ongoing life force. It is not necessarily the ultimate or the final expression. That is in the future. That next expression is the fodder for our expectation.

Finally, and most important, the resurrection is the breakthrough from self-consciousness to whole-consciousness. Jesus is the completion of human speciation. He is the prototype human. He is what it is to be a human. He is the whole-conscious human. He is the product of natural selection, the one who fits the niche harmoniously, pointing the way for the rest of humanity, calling the rest of humanity to follow him. Jesus was the one who expanded self-consciousness into whole-consciousness and called the rest of us to that consciousness, if we were to survive as a species.

The equation of the resurrection and the afterlife puts the emphasis in the wrong place. First and foremost, the resurrection has to do with life here on this Earth and the possibility for humans to expand beyond self-consciousness to whole-consciousness. To concentrate on life after death plays right into our tendency to ignore this world in favor of escape to a next, so-called higher, spirit world. It is the worst consequence of dualism. For ignoring this world in favor of an afterlife will lead inevitably to the end of human life on this Earth.

I have no doubt about the existence of something beyond physical death. There is too much scriptural evidence, not to mention actual human experience, to doubt its reality. But we humans have put our religious energies and imaginations all too much on the afterlife, and in doing so, we put our continued existence on the Earth in jeopardy.

The resurrection appearances were God's emphasis on the Earthly life-giving potential of Jesus and the Way he walked. Those appearances were the "glow" of this one human, living a healthy life for the whole. Jesus breached the walls of self-consciousness and led humanity into the beyond of whole-consciousness.

In terms of the Dynamics, whole-consciousness is the logical outcome of the becoming of being and belonging. We have seen that being progresses toward increasing consciousness and belonging moves toward increasing diversity. Whole-consciousness is the culmination of these tendencies in the human species. It is the union of consciousness and diversity. Whole-consciousness is finally human consciousness of the diversity of the whole.

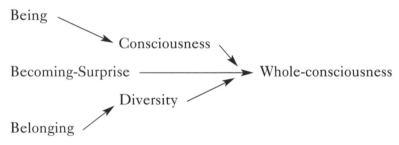

God appeared in Jesus in being and belonging, becoming and surprise, so that we humans could know how to live in whole-consciousness. Jesus introduced us to the community of all of life—from the simplest atom to the most complex mammal, from the most elemental particles of energy to the huge revolving galaxies. Jesus showed us the way to the whole. "I am the way, and the truth, and the life. No one comes to the Father except through me" (John 14:6).

JESUS, THE POWER FOR HUMAN SURVIVAL

Jesus not only shows the way, he is the power for humans to live in the universe. He was a man of humility and restraint with regards to himself. He protected the weak. He talked about sufficiency in terms of what was needed for today only. He championed justice for all.

Since Jesus was the incarnation of the very heart of the universe, the power of Jesus is the power of the universe. Walk in this way and you are in line with the universe. Walk in this way and you learn what it is to be a human being.

I can remember learning how to play baseball as a child. It was difficult and I was awkward. Then I went to Ebbets Field in Brooklyn to watch the Dodgers play. I was transfixed with Peewee Reese at shortstop. He was more than I had ever dreamed of. From then on, I moved on the field with a new grace. I had finally encountered the standard of what it is to play baseball.

Jesus came among us, lived, died, and was raised again to show us what it is to be a human being. In Jesus is the power that will live.

The spirit of Jesus is the spirit of the universe given in a form to which we can relate. As we let this spirit indwell us, we let the universe into ourselves. We are then in mind, spirit, and body part of the creation, part of the universe. No longer do we need to act as if we are separate from the creation. In Christ, we can become one with the creation again.

Now, life following Jesus is not one of guaranteed success for each individual organism. That is the rub for us self-conscious humans. We want life for ourselves individually, on our own terms. The life promised is life for the whole—for the whole human race, for the whole community of species, for the whole Earth. From this perspective, individual life and death is so much less important. It is just part of the ongoing life story.

If Jesus did anything, he showed us how, as individuals, to die gracefully. He showed us that in dying is life. That is the final hope of the resurrection. We can die in the faith that death is not the end. By surprise, death is the way to life. That is the deepest story of the universe.

Finally, Jesus showed us how to play. We have evolved from the infant chimpanzee. We are by heritage and destiny a playful species. Our big brain self-consciousness has driven us into the fear of death and a tenacious need to control. But we are born to be playful, loving, adoring, spontaneous, childlike. As Jesus said, "Truly I tell you, whoever does not receive the kingdom of God as a little child will never enter it" (Luke 18:17).

When my oldest son was about 18 months, one of the first objects he learned to name was a bus. We would be driving along through the city and he would yell at the top of his voice, "Bus, bus!" We would look around and then spot a city bus three blocks away, barely noticeable. His childlike delight was captivating, irrepressible, and of the essence of what it is to be human.

We are made to take delight in the creation around us. Classical books on prayer often name five kinds of prayer. The one that is at the top is always adoration. Again, we are made for wonder.

The physicist, Brian Swimme, says it most playfully. He says that humans are built to "gawk." We are made to be like country farmers come into the big city for the first time, our heads thrown back as we stare at the skyscrapers, our mouths hanging open in a giant gawk.

The creation is full of wonder. Everything in it is unique and unrepeatable. It is turbulent and organic. It is relative—of Einstein, not Descartes. We are not here to control it but to admire it. That is our central function as humans. Finally, the Earth has a species that can sing praise to the Creator. Finally, the Earth can rejoice in its creation.

When Jesus urged us to be childlike, he did not mean for us to be childish and irresponsible. He meant that we humans are to have a certain innocence that delights in what is and stays open to what is possible. We are to have a childlike trust that will rely on the creation for life. We are to be childlike, just like our chimpanzee forebears, just like Jesus, if we would survive. The spirit of the four Dynamics lives and can empower us, indwell us.

The Church in the Ecozoic Era

Finally, let us consider the coming place of the church. Here we don't just mean the organization that is called church to which duly baptized believers belong. Rather, we're speaking in a looser sense of the mystical church that is the company of all believers. It includes the Christian church, but the church we're speaking of also includes those who live as Christ lived, those who love God and care for others as they care for themselves. It is not so much a legal definition as a spiritual one.

We have traditionally said that the church is the body of which Christ is the head and all believers are the members. It is the people of God. The church, therefore, picks up where Jesus left off when he died and arose again. The church has the function in the present that Jesus had when he walked the Earth. It is not straining too much to say that the church is the extension of Jesus. The church is the Christ.

If the church is the Christ, and Christ is the four Dynamics, then in the terms we have been using, the church is being and belonging, becoming and surprise. The church, in its larger, mystical, ideal state, is the present, human embodiment of the Dynamics that have been from the beginning, the Dynamics that govern the universe. All those who live by those Dynamics are believers and members of the church.

Every species that has ever successfully speciated lived by the four Dynamics. They are then, in some sense, believers. They follow the Christ just as surely as any born-again Christian follows the Christ. From this point of view, every successful species then is a member of the church.

The true, mystical, universal church is, therefore, the company of all of life, organic and inorganic—stardust and galaxies, amoeba and polar bears, field mice and eagles, beta particles and lightning storms, Earth worms and physicists. The church is the whole universe arrayed around the heavenly throne of God, singing, dancing, and praising the One who empowers it.

In the becoming universe, every new day is a Eucharist in this church. And the Christ continually presides over this Eucharist of life, to the honor and glory of God. The altar is surrounded by acolytes and choirs. Those attendants are all of life. The buzz of the bees and the song of the birds are all canticles of praise. The gurgling of waters and the sigh of the wind are the wafting prayers of incense. The inevitable suffering of life as the universe becomes is the bread to be broken and the wine to be poured. The continual burstings forth of new life are the shouts of praise.

We humans are the newcomers, sitting in the back pews waiting to be invited forward to join the angelic chorus. We hardly know how to behave. We see through a glass darkly and are quite awkward. We barely believe.

The Christ who lived, died, and lived again in Jesus is our sponsor as we practice the rounds of faith—being, belonging, becoming, and surprise. Through him, we rehearse our being as creatures built to gawk. Through him, we exercise our belonging as members of a great community of life. Through him, we perform our becoming as those who live in tiptoe expectation of the future. Through him, we learn to live by surprise, dying gracefully that we might live for the whole creation. It is through Christ that we know what it is to live in this great universe.

At the center of everything is God. God powers the whole dance of the universe. God has been forever and yet is always new. We have no idea how this story will evolve. We can only trust that the God dedicated to life will prevail.

One thing we do know is that the story will unfold by surprise. And that surprise will be a bounteous gift. For the new day coming will be like scarlet ribbons for our beloved's hair. The surprise in that new day will be like the wonder of early morning frost on a winter's window. The surprise will appear suddenly, like a magnificent buck deer on the trail or the graceful flight of a great blue heron across the lake. The surprise of the becoming God will be there to greet us after this hard hike, just like those swooping, dancing eagles were there to greet Ed and me on that remarkable day at the Jacks River Falls.

Surprise!

ANNOTATED BIBLIOGRAPHY

This is not an exhaustive list of books that are related to the creation and a Christian understanding of it. This is, rather, a list of the books that I encountered on my path of discovery, a path that was limited by time and the constraints of full-time employment, but was driven by a passion to know and understand. I share them as jewels discovered along the way.

Ernest Becker, *The Denial of Death*, New York: The Free Press, 1973. Becker, in this book and its companion, *Escape From Evil*, makes a brilliant case for why humans behave as destructively as they do.

Ernest Becker, *Escape From Evil*, New York: The Free Press, 1975. In this sequel to *The Denial of Death*, Becker shows how social evils derive from the basic human situation described in his first volume.

Thomas Berry, *The Dream of the Earth*, San Fransisco: Sierra Club Books, 1988. Berry is the modern dean of those who reintegrate spirituality, the creation, and the role of humanity. This is the must-read primer on the subject.

Eric Chaisson, *The Life Era*, New York: The Atlantic Monthly Press, 1987. This book helped me understand some of the events in the early evolution of the universe.

Niles Eldredge, *Life Pulse*, New York: Facts On File Publications, 1987. This is the book that started my recent journey. It talks about evolution and mass extinctions and an emerging new refinement of evolutionary thought that is open to theological interpretation.

Niles Eldredge, *Time Frames*, New York: Simon & Schuster, 1985. The theory of punctuated equilibria is explained along with the thrilling detective story that led to the theory.

Niles Eldredge and Ian Tattersal, New York: *The Myths of Human Evolution*, Columbia University Press, 1982. Eldredge and Tattersal describe the scientific evidence pointing to the emergence of humanity into a full-blown species.

Matthew Fox, *Original Blessing: A Primer on Creation Spirituality*, Santa Fe, NM: Bear and Company, 1983. This book reveals one of the deep intellectual fault lines in Western thinking that has lead to the dominance of humans over the creation.

James Gleick, *Chaos: Making a New Science*, New York: Penguin, 1987. This is a good book to read if you want to be assured of the end of the Newtonian cause-and-effect world.

Stephen Jay Gould, *Ontogeny and Phylogeny*, Cambridge, MA: The Belknap Press of Harvard University Press, 1977. This is a detailed scientific work that describes a theory on the evolutionary quirk that produced humanity from the chimpanzee.

John F. Haught, *Science and Religion*, New York/Mahwah, NJ: Paulist Press, 1995. This book presents a thorough outline of the various positions taken in the current dialogue between science and religion. The author clearly comes down on the side of an expanding resonance between the two.

Stephen Hawking, *A Brief History of Time*, New York: Bantam Books, 1988. *A Brief History of Time* has been described as one of the most purchased and least read books of our time. It is tough sledding but, for the persistent reader, it will describe the wonders of modern scientific thinking.

James Lovelock, *The Ages of Gaia*, New York: Bantam Books, 1988. Far from being a New Age book, this popularized scientific work opens the possibility of seeing the Earth as a dynamic, living system.

Sallie McFague, *Models of God*, Philadelphia: Fortress Press, 1988. Written from a feminist perspective, this is an important work searching for a new metaphor for the relationship between God and the creation.

Peter Russell, *The Global Brain*, Los Angeles: J. P. Tarcher, Inc., 1983. Russell provides some interesting speculation on evolution and human consciousness.

Elisabet Sahtouris, *Gaia, The Human Journey from Chaos to Cosmos*, New York: Pocket Books, 1989. This small book chronicles the remarkable evolutionary journey of life from the beginning up to humanity. One stands in awe of the procession after reading this book.

Barbara Brown Taylor, *The Luminous Web*, Cambridge, MA: Cowley Publications, 2000. Taylor is one of the best preachers in America today and this, sometimes poetical, often profound, inquiry into the boundary land between science and religion will produce awe in your soul.

Pierre Teilhard de Chardin, *The Divine Milieu*, New York: Harper Torch-
     books, 1960. Teilhard was a scientist and a mystic. He explores the
     mystical development of humanity in this small book.

Pierre Teilhard de Chardin, *The Phenomenon of Man*, New York: Harper
     Torchbooks, 1959. This is Teilhard's seminal work, describing a new
     way of looking at humanity and the creation as one and on one jour-
     ney. This book was pivotal in my early thinking.

Edward O. Wilson, *Consilience*, New York: Alfred A. Knopf, 1998. Written
     by a leading scientist, *Consilience* is a thorough exploration of the pro-
     found need for all of the specialized disciplines of science and the
     humanities to converge on a search for the truth in order to preserve
     the world from environmental disaster.

World Commission on Environment and Development, *Our Common
     Future*, New York: Oxford University Press, 1987. The final report by
     an official United Nations' commission, chaired by Gro Harlem
     Brundtland of Norway, compellingly ties together the existence of
     poverty and environmental degradation.

The Lineage of Humanity

Creation Event	Time
The creation	15 billion years ago (b.y.a.)
The first hydrogen	1 million years after the beginning
The first galaxies	1 billion years after the beginning
Stars, red giants, white dwarfs, and supernovas	Continuously
The birth of the sun	5 b.y.a.
The formation of the Earth	4.5 b.y.a.
Self-reproducing life (microbes)	3.75 b.y.a.
Fermenters	1 to 3 b.y.a.
Photosynthesizers	1 to 3 b.y.a.
Breathers	1 to 3 b.y.a.
Multicelled creatures	1 b.y.a.
DNA programmed death	800 million years ago (m.y.a.)
Shells and skeletons	540-550 m.y.a.
The first plants on land	425 m.y.a.
Creatures on land	395 m.y.a.
Trees	370 m.y.a.
Amphibians	370 m.y.a.
Reptiles	313 m.y.a.
Dinosaurs	235 m.y.a.
Mammals	220 m.y.a.
Birds	150 m.y.a.
Flowers	110 m.y.a.
Primates	70 m.y.a.
The great extinction that took dinosaurs	65 m.y.a.
Hominids	5 m.y.a.
The first modern humans— *Homo sapiens*	40,000 years ago
Recorded history	5,500 years ago

APPENDIX 2

Where to Go for Current Environmental Information

For general information and reports

United Nations Environmental Programme (UNEP)
P.O. Box 30552
Nairobi, Kenya
Phone: (254 2) 621234; Fax: (254 2) 226886 / 622615
www.unep.org

United Nations Population Information Network (POPIN)
The Director, Population Division
Department of Economic and Social Affairs
United Nations Secretariat
2 United Nations Plaza (Rm. DC2-1950)
New York, New York 10017, USA
Phone: (212) 963-3179; Fax: (212) 963-2147
www.un.org/popin

World Resources Institute
10 G Street NE (Suite 800)
Washington, DC 20002, USA
Phone: (202) 729-7600; Fax: (202) 729-7610
www.wri.org

For specific issues and action suggestions

Earthwatch Institute
3 Clock Tower Place, Suite 100
Box 75
Maynard, MA 01754, USA
Phone: 1-800-776-0188; Fax: (978) 461-2332
www.earthwatch.org

Environmental Defense Fund
257 Park Avenue South
New York, New York 10010, USA
Phone: (212) 505-2100; Fax: (212) 505-2375
www.edf.org

National Wildlife Federation
11100 Wildlife Center Drive
Reston, VA 20190-5362, USA
Phone: (703) 438-6000
www.nwf.org

Sierra Club
85 Second St., Second Floor
San Francisco, CA 94105-3441, USA
Phone: (415) 977-5500; Fax: (415) 977-5799
www.sierraclub.org

Wildlife Conservation Society
2300 Southern Boulevard
Bronx, New York 10460, USA
Phone: (718) 220-5100
www.wcs.org

World Wildlife Fund
www.wwf.org

INDEX

Air, current situation, 113–16
Aquinas, Thomas, 37

Bacon, Francis, 17–18, 20
Bacteria, bluegreens (photosynthesis), 59–60, 89; breathers
   (respiration), 59, 89; bubblers, 58–59, 89
Baptism, 24
Becker, Ernest, 91
Becoming Dynamic, definition
   of, 27; in early universe,
   35–36; in Ecozoic Era,
   141–45; in essence of God,
   38; in evolution, 69; in Gaia,
   66
Being Dynamic, and Belonging
   Dynamic, 48–49; definition
   of, 26; in Ecozoic Era,
   133–35; and Gaia, 65; and
   quantum mechanics, 45–46
Belonging Dynamic, and Being
   Dynamic, 48–49; definition
   of, 26–27; in early seas, 61; in
   Ecozoic Era, 136–41
Berry, Thomas, 3, 16, 94, 106,
   132–33; principles of, 24–25

Black Death (the plague), 16,
   18, 20
Brundtland, Gro Harlem, 110,
   137
Buber, Martin, 136

Calvin, John, 17–20
Cape Hatteras, 81–82, 89–90, 95
Cenozoic Age, 80, 133
CFCs, 8, 115, 117
Chaos theory, 46, 141, 145
Chief Seattle, 7
Church, 19, 23, 29; in Ecozoic
   Era, 154–56; and Four
   Dynamics, 120–27; task for,
   130
Complexity, as cosmological
   principle, 23–27; in early seas,
   54–58, 60–62; in early universe, 32, 35–36; in evolution,
   65–68, 75–76, 144; of humans,
   141, 147; present, 109
Consciousness, collective, 148;
   in cosmos, 24, 26–27, 35, 142;
   definition of, 45–46, 88;
   dimension of being, 45–46; in
   early seas, 57–58; progression

Consciousness *(continued)*
toward, 65–67, 75, 141, 144,
146, 152
Constable, John, 135
Copernicus, 46
Cosmology, in culture, 9–11, 28;
domination, 18–20; dualistic,
22; emerging, 128–29; integrated, 16, 19–20; medieval,
36–37; Native American,
11–12, 143; new, 23–25,
28–29, 39, 133, 135, 141,
145; and present situation,
117; religious, 16, 19–20,
140; Western, 12–13, 16, 18,
23
Council of All Beings, 139
Creationism, 77
Cro-Magnon man, 85

Darwin, Charles, 67–68
Death, in cosmology, 9; fear and
anxiety about, 91–92, 127,
153; invention of, 62; in
Native American cosmology,
12–13; part of surprise,
33–34, 102, 145; and resurrection, 100, 149; in Western
cosmology, 13
Descartes, René, 17–18, 20, 22,
56, 154
Diversity, in belongings, 26;
part of becoming, 35; present, 109; progression toward,
65–67, 75–76, 144; and sustainability, 136–37; and
whole consciousness, 152

Dominion over the creation,
13–14, 129
Dualism, dangers in, 56, 58,
100, 119, 125, 141, 143, 151;
in image of God, 39; in new
cosmology, 129; problem of,
22–23
Dynamics, the Four, and
church, 120–30; in evolution,
75–76; and Gaia, 64–66; and
Jesus, 100–103; outline of
25–28; in present, 117–20;
and science, 78; *see also*
Becoming Dynamic; Being
Dynamic; Belonging
Dynamic; Surprise Dynamic

Earth Summit, 8
Ecozoic Era, 133; and becoming, 141–45; and being,
133–35; and belonging,
136–41; and church, 154–56;
and surprise, 145–46
Einstein, Albert, 8; and benign
universe, 34; general agree-
ment with, 128–29; and new
cosmology, 141, 145; Theory
of Relativity, 37, 41–44
Eldredge, Niles, 68, 71
Extinction, exhibit on, 6; great,
73–74, 110; idea of before
Darwin, 67; rate of, 6, 109;
replace disease and famine;
107
Evolution, 67; punctuated equilibria, 69–71; retarded development of humans, 86–87, 94

Fall, the, 13–4, 97–98
Fossils, 63, 67, 84
Francis of Assisi, 15–16, 20

Gaia, 115, 120–21, 129; theory
    of, 63–64
Galileo, 46
Genesis, 13–14
Global warming, 8, 111–14,
    133
Gould, Stephen Jay, 68, 71,
    85–86, 93
Greek heritage, 15
Greenhouse gases, 7, 112–14,
    117

Handel, George Frederick, 36
Haught, John F., 76
Hawking, Stephen, 41
Heisenberg, Werner, develop-
    ment of, 44–45; in relation to
    Newtonian mechanics, 48;
    uncertainty principle, 37
Holy Eucharist (Communion),
    125, 155
Homo sapiens, abilities of, 89;
    with being and belonging,
    117–18; cause of extinction,
    6; childlikeness in, 87,
    153–54; and control, 93;
    emergence of, 73, 75, 83–
    85; and Jesus, 103; and spirit,
    98
Hubble, Edwin, 41–42

Jacks River Falls, 131, 156
Jesus, the Christ, 148–56; and

church, 120–22; and Ecozoic
    church, 154–56; place in Four
    Dynamics, 25, 28–29, 100–
    103, 143, 145–46; power for
    survival, 152–54; resurrection
    of, 149–52; and theology,
    127–29; and whole conscious-
    ness, 148
Justice, human, 121; and Jesus,
    152; and various species,
    137–38

Khruschev, Nikita, 38

Land, current situation, 110–12
Love Canal, 104, 116, 129
Lovelock, James, 64

Manhattan Project, 95
Martin, Danny, 94
McFague, Sallie, 39
McGillis, Miriam, 9
Mitchell, Edgar, 99
Montreal Protocol, 8, 115

Native Americans, 11–12, 124,
    143
Neanderthal Man, 85, 88
Newton, Sir Isaac, 37; and
    belonging, 48–49; Newtonian
    physics, 44, 123; and relativ-
    ity, 41–42, 45–46

Ozone layer, 7, 115

Paul, St., 149
Planck, Max, 44

Population, growth, 107–8; and consumption, 116
Poverty, 110–12, 116

Quantum mechanics, 41, development of theory of, 44–45; and new cosmology, 141, 145

Reilly, William, 136
Relativity, Theory of, *see* Einstein
Resurrection, death and, 100, 128; of Jesus, 145, 150–53; within universe, 33
Russell, Peter, 147–48

Self-consciousness, in humankind, 26–27, 87–93, 95, 117, 134, 142, 153; universe becoming, 146; and whole consciousness, 146–48, 151–52
Sin and evil, emphasized in religious thought, 16, 20; against humanity, 100; in humanity, 95–98, 118, 122
Speciation, definition of, 71; evil at time of, 98; and Four Dynamics, 155; humanity and, 118, 125, 128, 145–46
Spencer, Herbert, 68

Spirit, 98–100, 151; at beginning, 141–44
Surprise, Dynamic, definition of, 27–28; in early universe, 33–35; in Ecozoic Era, 145–46; in essence of God, 38; in Jesus, 149–54
Subjectivity, within Being, 24–25; in dynamics, 26–27; in universe, 33, 46
Swimme, Brian, 33, 154

Taylor, Barbara Brown, 78
Teilhard de Chardin, Pierre, 23–25
Theology, feminist, 129; of God, 36–39; liberation, 39, 129; process, 39, 129; reflection on cosmology, 28–29; and resurrection, 150; task for, 127–30
Tillich, Paul, 38–39
Turner, William, 135

Water, current situation, 112–13
Way of the Cross, 149–52
Whole consciousness, 146–48, 151–52
Wilbur, Ken, 94
Wilson, Edward O., 78